SpringerBriefs in Immunology

T0185091

For further volumes:
http://www.springer.com/series/10916

Anton Kutikhin • Elena Brusina
Arseniy E. Yuzhalin

Viruses and Atherosclerosis

Anton Kutikhin
Epidemiology
Kemerovo State Medical Academy
Kemerovo, Russia

Elena Brusina
Epidemiology
Kemerovo State Medical Academy
Kemerovo, Russia

Arseniy E. Yuzhalin
Res Inst Cardiovascular Diseases
Russian Academy of Medical Sciences
Siberian Branch
Kemerovo, Russia

Department of Oncology
University of Oxford
Oxford, UK

ISSN 2194-2773
ISBN 978-1-4614-8862-0
DOI 10.1007/978-1-4614-8863-7
Springer New York Heidelberg Dordrecht London

ISSN 2194-2781(electronic)
ISBN 978-1-4614-8863-7 (eBook)

Library of Congress Control Number: 2013949161

Printed on acid-free paper

Springer is part of Springer Science+Business Media (www.springer.com)

I dedicate this book to my dear Sigrid, who has been supporting me during the hardest year of my life, and who is still one of my best friends for over 9 years since that time. All the time I know that despite any circumstances, she is with me incessantly.

Preface

Over the years of investigations a lot of discoveries supporting the hypothesis of infection-driven atherosclerosis were made, and many associations between various infectious agents and atherosclerosis-related diseases were revealed. The list of these infectious agents includes *Chlamydia spp.*, *Helicobacter pylori*, *Mycoplasma pneumoniae*, *Streptococcus pneumoniae*, *Enterobacter hormaechei*, *Borrelia burgdorferi*, periodontal microorganisms, human immunodeficiency virus, cytomegalovirus, influenza viruses, hepatitis A, B, and C viruses, herpes simplex virus-1 and -2, Epstein-Barr virus, enteroviruses, parvovirus B19, respiratory syncytial virus, and measles virus. The problem is of large importance due to the enormous worldwide incidence, prevalence, and mortality from atherosclerosis-related diseases. To the best of our knowledge, this is the only book devoted to this problem and highlighting its importance and topicality. However, the amount of published information about infection-driven atherosclerosis is extremely large, so here we have focused only on viruses but not on bacteria. In the first chapter, we describe the hypothesis and denote the mechanisms of virus-driven atherosclerosis. Further in the book, the role of various viruses in the development of atherosclerosis and atherosclerosis-related diseases is analyzed; we start from herpes simplex virus-1 and -2, move to Epstein-Barr virus, continue with enteroviruses, parvovirus B19, respiratory syncytial virus, measles virus, and finish by the hepatitis A, B, and C viruses. In the last chapter, we write about the practical applications of the basic and epidemiological investigations performed in the field. Concerning other viruses, the role of cytomegalovirus and influenza virus in the development of atherosclerosis and related diseases was recently described in comprehensive reviews, and we do not want to repeat the information instead of carrying out the analysis. So, we indicate the relevant references to these papers, and everyone who wants to acquaint with the problem can find them in the Web or in libraries. We apologize for this inconvenience, but the goal of this book is to perform the analysis of published data but not to repeat the information that was written before in the review articles.

We prepared this book with the goal that it will be useful for the wide audience, particularly cardiologists; angiologists; microbiologists; epidemiologists; PhD, graduate, and undergraduate students of biomedical faculties; and their lecturers.

Kemerovo, Russia   Anton Kutikhin
                                                                                                   Elena Brusina
                                                                                           Arseniy E. Yuzhalin

Contents

Chapter 1
A Hypothesis of Virus-Driven Atherosclerosis

Abstract In this chapter, we will briefly describe a hypothesis of virus-driven atherosclerosis. First, we will write about the pathogenetic basis of atherosclerosis, then we will note the possible place of infectious agents in this pathological process, and, finally, we will consider a number of mechanisms which underlie the hypothesis of virus-driven atherosclerosis.

Atherosclerosis, manifesting itself as acute coronary syndrome, stroke, and peripheral arterial diseases, is a chronic progressive inflammatory disease characterized by the accumulation of lipid and fibrous elements in arterial walls [1], which is driven by responses of both innate and adaptive immunity [2–4]. Pathogenically, it starts from the deposition of cholesterol in the subendothelial region of the vessel under the condition of local endothelial dysfunction and hypercholesterolemia and/or dyslipidemia [5]. This triggers the infiltration of leukocytes, monocytes/macrophages, T cells, mast cells, neutrophils [6, 7], and dendritic cells [8, 9], causing an asymmetrical thickening of the intima. The continuous influx of cells and permanently developing immune response and inflammation (driven by reactive oxygen species, cytokines, complement factors and proteases) maintain progression of atherosclerotic plaque, and lead to the formation of complex mature plaques [10, 11]. They consist of a necrotic core formed by apoptotic, necrotic cells, cholesterol crystals, and cell debris, and they are surrounded by the immune cells indicated above. Additionally, mature plaques are covered by smooth muscle cells and a collagen-rich extracellular matrix [3, 4]. The release of proinflammatory cytokines and proteases may break the collagen fibers and disrupt the fibrous cap, resulting in plaque rupture and acute thrombotic occlusion of the vessel [2–4, 10–12]. The underlying mechanism of the chronic inflammatory process in atherosclerosis is still unknown in a significant extent. As a possible trigger, several studies have suggested that various bacteria and viruses are associated with atherosclerotic diseases, the most clinically significant of which are stroke and coronary artery disease [13–15]. The list of such infectious agents may include *Chlamydia spp.*, *Helicobacter pylori*,

Mycoplasma pneumoniae, *Streptococcus pneumoniae*, *Enterobacter hormaechei*, *Borrelia burgdorferi*, periodontal microorganisms, cytomegalovirus, Epstein-Barr virus, hepatitis B and C viruses, herpes simplex virus-1 and -2, enteroviruses, parvovirus B19, respiratory syncytial virus, measles virus, influenza virus, and possibly a number of other microorganisms [16].

Certain mechanisms by which viruses may cause atherosclerosis and related diseases can be suggested:

- Immune-mediated mechanisms performing via either direct or indirect inflammation. Direct inflammation can be caused by virus infecting endothelial cells or immune cells attracted to the site of endothelial dysfunction (for instance, lymphocytes), with further initiation and promotion of atherosclerotic plaque formation. Indirect inflammation may be caused by virus infecting every other part of the human body (for instance, liver), with further elevated production of cytokines and acute phase proteins, which may potentiate the inflammation in blood vessels promoting the development of atherosclerotic plaques. In addition, viruses may also cause autoimmune conditions via the phenomenon of crossreactivity; their antigens may cause immune reaction against the similarly structured antigens of the host tissues and organs, that may also create the condition of indirect chronic inflammation;
- Atherotic mechanisms performing via changing of the lipid profile from favorable/normal to unfavorable due to affection of liver or to other reasons. Unfavorable lipid profile (high level of total cholesterol, triglycerides and low-density lipoprotein cholesterol, low level of high-density lipoprotein cholesterol) increase the risk of atherosclerosis development itself, initiating and promoting plaque formation;
- Coagulative mechanisms, which are performed by the alteration of coagulative ability, leading either to atherothrombosis (if it is increased) or to hemorrhage possibly causing hemorrhagic stroke (if it is decreased). An ability to coagulation may be altered, for instance, due to liver affection;
- In addition, viruses may alter gene expression in the infected cells, feasibly causing higher production of substances promoting plaque formation (for instance, growth factors or modifiers of extracellular matrix).

In the following chapters, we will consider the role of various viruses in the development of atherosclerosis and related diseases (for instance, coronary artery disease, myocardial infarction, and stroke). We will also describe the mechanisms via which these viruses perform their hazardous activity.

References

1. Hansson GK. Mechanisms of disease: inflammation, atherosclerosis, and coronary artery disease. N Engl J Med. 2005;352(16):1685–95.
2. Hansson GK, Robertson AK, Soderberg-Naucler C. Inflammation and atherosclerosis. Annu Rev Pathol. 2006;1:297–329.

3. Galkina E, Ley K. Immune and inflammatory mechanisms of atherosclerosis. Annu Rev Immunol. 2009;27:165–97.
4. Weber C, Noels H. Atherosclerosis: current pathogenesis and therapeutic options. Nat Med. 2011;17(11):1410–22.
5. Steinberg D. Atherogenesis in perspective: hypercholesterolemia and inflammation as partners in crime. Nat Med. 2002;8(11):1211–7.
6. Zernecke A, Bot I, Djalali-Talab Y, Shagdarsuren E, Bidzhekov K, Meiler S, et al. Protective role of CXC receptor 4/CXC ligand 12 unveils the importance of neutrophils in atherosclerosis. Circ Res. 2008;102(2):209–17.
7. Drechsler M, Megens RT, van Zandvoort M, Weber C, Soehnlein O. Hyperlipidemia-triggered neutrophilia promotes early atherosclerosis. Circulation. 2010;122(18):1837–45.
8. Manthey HD, Zernecke A. Dendritic cells in atherosclerosis: functions in immune regulation and beyond. Thromb Haemost. 2011;106(5):772–8.
9. Weber C, Meiler S, Döring Y, Koch M, Drechsler M, Megens RT, et al. CCL17-expressing dendritic cells drive atherosclerosis by restraining regulatory T cell homeostasis in mice. J Clin Invest. 2011;121(7):2898–910.
10. Ross R. The pathogenesis of atherosclerosis: a perspective for the 1990s. Nature. 1993;362(6423):801–9.
11. Libby P. Inflammation in atherosclerosis. Nature. 2002;420(6917):868–74.
12. Weber C, Zernecke A, Libby P. The multifaceted contributions of leukocyte subsets to atherosclerosis: lessons from mouse models. Nat Rev Immunol. 2008;8(10):802–15.
13. Benditt EP, Barrett T, McDougall JK. Viruses in the etiology of atherosclerosis. Proc Natl Acad Sci U S A. 1983;80:6386–9.
14. Valtonen VV. Infection as a risk factor for infarction and atherosclerosis. Ann Med. 1991;23(5):539–43.
15. Xu Q, Willeit J, Marosi M, Kleindienst R, Oberhollenzer F, Kiechl S, Stulnig T, Luef G, Wick G. Association of serum antibodies to heat-shock protein 65 with carotid atherosclerosis. Lancet. 1993;341(8840):255–9.
16. Rosenfeld ME, Campbell LA. Pathogens and atherosclerosis: update on the potential contribution of multiple infectious organisms to the pathogenesis of atherosclerosis. Thromb Haemost. 2011;106(5):858–67.

Chapter 2
The Role of Herpes Simplex Virus-1 and Herpes Simplex Virus-2 in Atherosclerosis

Abstract In this chapter, we discuss a relationship between herpes simplex virus (HSV) and atherosclerosis. According to the current reports, there are several pathways which may lead to the development of HSV-driven atherogenesis. Among these pathways are alteration of lipid metabolism, prothrombic activity, and expression of lectin-like oxLDL receptor (LOX-1). In recent decades, multiple epidemiological studies have been conducted in order to reveal HSV in plaques or in serum of atherosclerotic patients. In addition, a plenty of case-control studies has been performed to reveal potential correlation between the presence of HSV and a risk of atherosclerosis and related diseases. After analysis of numerous studies we suggest that HSV could hardly be the sole causative agent of atherosclerosis; however, its role may be important as a part of pathogen burden (five or more pathogens).

2.1 Is There a Link Between HSV and Atherosclerosis?

Herpes simplex virus (HSV) is an ubiquitous pathogen which belongs to herpes virus family, *Herpesviridae* [1]. The virus exists in two forms, HSV-1 and HSV-2, which share 50 % homology in DNA sequence [1]. HSV-1 causes fever blisters and sore cold, and is transmitted through oral contacts, whereas HSV-2 is responsible for genital herpes and can be transmitted only sexually. Importantly, both HSV-1 and HSV-2 are highly resistant and cannot be fully eradicated from the body; therefore, worldwide HSV prevalence is approximately 65–90 % in accordance with the latest reports [2]. Today, there is substantial basis to suppose that HSV can contribute to the occurrence and development of atherosclerosis and related cardiovascular diseases in risk groups (we discuss them below). Multiple studies have been performed in order to clarify whether HSV-1 and HSV-2 are causative agents of atherosclerosis; however, these investigations are not structured and scattered. In this chapter we summarize and discuss existing findings revealing possible correlation between HSV, atherosclerosis and related diseases.

A. Kutikhin et al., *Viruses and Atherosclerosis*, SpringerBriefs in Immunology 4,
DOI 10.1007/978-1-4614-8863-7_2, © Springer Science+Business Media New York 2013

In 1975, Blacklow et al. [3] discovered that human aorta cultures support viability and replication of certain viruses, including HSV-1. This curious observation initiated further in-depth research in order to elucidate the role of this virus in the etiopathogenesis of atherosclerosis. In 1983, Benditt et al. [4] found herpes simplex viral mRNA in arterial plaque cells of patients undergoing coronary bypass surgery. The authors hypothesized that the presence of HSV in the arterial wall may lead to intimal hyperplasia, and consequently promote development of atherosclerosis. Later, the studies by Minick et al. [5] and Hajjar et al. [6] independently demonstrated that specific-pathogen-free chickens with Marek's disease herpesvirus (MDV, a pathogen that is very similar to HSV) develop chronic atherosclerosis like that in humans. The authors suggested that MDV altered metabolism of lipids, particularly, cholesteryl esters and cholesterol, which consequently led to lipid accretion and atherosclerosis. Furthermore, enhanced cholesterol and cholesteryl ester accumulation was observed in cultured avian arterial smooth muscle cells (SMC), according to the findings by Fabricant et al. [7] and Hajjar et al. [8]. In their further investigation, Hajjar et al. [9] confirmed that HSV-1 is able to induce lipid accumulation in human fetal and bovine adult arterial SMC similar, in part, to the lipid accumulation observed in vivo during human atherogenesis. Moreover, Visser et al. [10] demonstrated that HSV-infected human umbilical vein endothelial cells possess prothrombic activity due to enhanced assembly of prothrombinase complex on their surface. Proceeding from this fact, it could be suggested that HSV may lead to activation of massive coagulation, platelet adherence, and, consequently, atherogenesis. Additionally, Key et al. [11] revealed that HSV-infected cells have a loss of surface thrombomodulin activity and an increased synthesis of tissue factor, which also can promote atherosclerotic thrombosis. It is also necessary to mention an interesting investigation by Etingin et al. [12], who proposed the model that HSV induces endothelial cells surface expression of HSV glycoprotein C, which is a binding site for prothrombinase. The authors state that this leads to the generation of thrombin which, in turn, may activate platelets and subsequently result in cell injury, inflammation, and atherogenesis. Recently, Chirathaworn et al. [13] revealed that HSV infection can lead to the increase expression of lectin-like oxLDL receptor (LOX-1) in endothelial cells. Importantly, LOX-1 has been reported as a potential trigger of early atherosclerosis due to CD68 expression (1), p38 MAPK activation (2), and decrease in interleukin 10 and superoxide dismutase expression (3) [14–16]. Futhermore, LOX-1 is considered as a critical determinant of reactive oxygen species (ROS) formation, inflammation, and endothelial dysfunction [14–16]. So, this may be yet another pathway of HSV-mediated atherogenesis. Finally, according to the results by Span et al. [17], infection of endothelial cell monolayers with HSV-1 results in an increased monocyte and polymorphonuclear leukocyte adherence, which also may indicate the role of this virus in atherosclerosis. Thereby, all the above-mentioned findings strongly support the hypothesis that HSV may be a causative agent of atherosclerosis, and consequently, lead to the occurrence and development of various cardiovascular diseases (CVDs).

2.2 Evidence from Epidemiological Studies

When it has become clear that HSV can contribute to atherosclerosis, a plethora of epidemiological studies have been conducted. In 1984, Gyorkey et al. [18] detected virions of the *Herpesviridae* family in SMC and rare endothelial cells in proximal aorta of 10 atherosclerotic patients out of 60. A series of studies by Yamashiroya et al. [19, 20] revealed presence of HSV-2 antigen and genome in coronary arteries and the thoracic aorta of 8 of 20 subjects without any CVD, and in coronary arteries of 2 of 4 patients with severe atherosclerosis. Importantly, the majority of virus-positive cases had early or advanced atheromatous changes, including lymphocytic infiltrates [19]. In 1992, Jackel et al. [21] examined endomyocardial biopsy specimens from 29 German patients who received an orthotopic heart transplant, and revealed that herpesviral nucleic acids were detected in 25 % of all biopsies. Sorlie et al. [22] conducted a large-scale case-control study, which included 340 patients with early atherosclerosis (thickened carotid artery) and 340 matched controls without a history of CVD. The authors found that the odds ratios for atherosclerosis risk were 1.41 ($P=0.07$) and 0.91 ($P=0.63$) for HSV-1 and HSV-2, respectively. When adjustment for potential confounders was performed, the odds ratios were 1.21 for HSV-1 ($P=0.45$), and 0.61 ($P=0.5$) for HSV-2. Although the results did not reach statistical significance, these findings suggest that presence of HSV-2 appears protective in relation to atherosclerosis risk. At the same time, possession of HSV-1 was associated with a 20 % increased atherosclerosis risk. Thereby, according to the results by Sorlie et al. [22], HSV-1 and HSV-2 have different impacts on the risk of occurrence of atherosclerosis for some unknown reasons. In contrast to these data, Raza-Ahmad et al. [23] demonstrated that 45 % of patients undergoing coronary artery bypass grafting were positive for antigen to HSV-2 and only one to HSV-1, which may possibly indicate the greater role of HSV-2 in the atherogenesis. However, it is necessary to note that only 31 subjects were enrolled in this study, and therefore these results should be interpreted with caution. Chiu et al. [24] demonstrated that HSV-1 serum antibodies were presented in 8 of 76 patients with significant carotid artery stenosis (>60 % narrowing of the lumen). Notable, the authors did not investigate prevalence of HSV-2 in this study. Interesting results were achieved by Espinola-Klein et al. [25], who enrolled 504 subjects suffered from various CVDs to measure intima-media thickness (IMT) and the elastic pressure modulus as markers of early atherosclerosis. After evaluation of prevalence of HSV types 1 and 2 in the study group, the authors found that there was a significant elevation of IgG antibodies against HSV-2 in subjects with increased carotid IMT ($P<0.0001$). Moreover, the authors indicated a significant increase in IgG antibody levels against HSV-2 in patients with advanced carotid atherosclerosis, compared to healthy individuals ($P<0.04$). However, these data were no longer significant after adjustment for several confounders, such as age, gender, coronary artery disease, peripheral artery diseases, and cardiac risk factors. Additionally, the authors failed to find any associations regarding the HSV-1. It is important to note that further investigations by Espinola-Klein et al. [26] revealed that presence of HSV-2, but not

HSV-1, is strongly associated with advanced atherosclerosis ($P < 0.02$). This effect was more evident after adjustment for age, sex, and cardiovascular risk factors [26]. Notably, the results by Espinola-Klein et al. [26] are consistent with findings by Raza-Ahmad et al. [23], suggesting a greater impact of HSV-2 in the development of atherosclerosis. However, in contrast to these studies, Watt et al. [27] failed to detect the presence of HSV-2 in carotid artery of 18 atherosclerotic patients, and the DNA of HSV-1 was identified only in 3 of 18 samples. Shi and Tokunaga [28] examined the presence of HSV-1 in 10 Japanese patients with atherosclerotic aorta and in 23 patients with non-atherosclerotic aorta. The authors reported that non-atherosclerotic tissues were tested positive for HSV-1 in 13 % (three) of subjects, and atherosclerotic aortic tissue was tested positive for HSV-1 in 80 % (eight). Ibrahim et al. [29] analyzed viral DNA of HSV-1 in 48 biopsies from atherosclerotic plaques extracted by end-arterectomy and in tissue from non-atherosclerosis vessels from 66 controls. The authors found that HSV-1 DNA was detected significantly more frequently in plaques (35 %) than in control veins (9 %, $P = 0.006$). Nevertheless, the frequency of HSV-1 DNA detection in the internal mammary artery grafts was as high as in plaques (22 %, $P = 0.28$). Importantly, the authors failed to reveal the presence of HSV-2 neither in cases nor in controls. Furthermore, Muller et al. [30] analyzed serum antibody titers of 109 consecutive German patients, who underwent surgery for high-grade internal carotid artery stenosis. Although seropositivity for HSV was found in 97.1 % of patients, plaque-PCR revealed viral DNA only in two subjects (3.9 %).

Interestingly, some data show that HSV can be prevalent in speciemens collected from healthy donors. Reszka et al. [31] reported that HSV DNA was found in 27 of 40 (67.5 %) and 17 of 20 (85 %) aortic wall specimens from Polish atherosclerotic patients and healthy controls, respectively. It is also noteworthy to mention a series of studies showing negative results. Kwon et al. [32] failed to detect HSV either in carotid artery of 128 Korean patients with atherosclerosis or in carotid artery of 20 normal arterial wall samples, obtained from transplant donors with no history of CVDs. Similarly, Skowasch et al. [33] revealed HSV neither in carotid artery of 21 German individuals with symptomatic in-stent restenosis nor in carotid artery of 20 subjects with stable angina. Voorend et al. [34] was unable to establish the presence of HSV-1 DNA in large vessel samples of 19 Dutch patients who died from various causes, but who all had signs of atherosclerosis in at least one of the cerebral arteries, varying from intima fibrosis to fully developed atherosclerotic plaques with calcifications. According to the findings by Hagiwara et al. [35], only 2 Japanese patients out of 50 had HSV DNA in atherosclerotic plaques extracted from carotid artery. Tremolada et al. [36] reported that genomic sequence of HSV-1 was not found in atherosclerotic plaques from carotid arteries obtained from 17 Italian patients. Altun et al. [37] did not detect positivity for nucleic acids of either HSV-1 or HSV-2 neither in coronary arteries extracted from 28 Turkish patients suffering from atherosclerosis nor in vein samples of 22 control subjects of the same ethnicity. Finally, no statistically significant relationships were obtained between the presence of HSV and either carotid intima-media thickness or coronary calcium prevalence in a large-scale population-based cohort study by Szklo et al. [38]

including 1,056 mixed US patients who were free of prevalent CVD. Moreover, even multiple adjustments did not change these patterns, although 85 % of enrolled subjects were positive for HSV.

2.3   Evidence from Cardiovascular Events

Apart from epidemiological studies concerning with determination of viral DNA in patients at different stages of atherosclerosis, there is accumulating evidence from case-control studies focused on detection of HSV in subjects who underwent cardiovascular events, such as stroke, myocardial infarction, etc. Negative results were reported by Ridker et al. [39], who conducted a large-scale, prospective study to determine the baseline prevalence of antibodies directed against HSV in 643 US men who subsequently developed a first MI or thromboembolic stroke and among 643 age- and smoking-matched men who remained free of reported vascular disease over a 12-year follow-up period. The authors found no evidence of any positive association for HSV prevalence, as the seropositivity was almost identical for cases and controls (69.0 % and 69.4 %, respectively). Moreover, the risk estimates were not substantially changed in multivariate analyses controlling for baseline differences in body mass index, blood pressure, hyperlipidemia, diabetes, or family history of premature atherosclerosis. Stratification by smoking status and the presence or absence of other cardiovascular risk factors also did not change the results significantly. In contrast to this study, Roivainen et al. [40] demonstrated that high level of antibodies against HSV-1 is a risk factor for future coronary events in a prospective cohort of 241 Finnish middle-aged dyslipidemic men who undergone either myocardial infarction or coronary death during the 8.5-year period of time. The authors reported that high antibody levels were associated with a two times increased risk of cardiovascular event in comparison with those possessing low antibody levels. Moreover, simultaneous occurrence of high C-reactive protein (CRP) and high HSV-1 levels resulted in 25-fold increased risk. As CRP is a mediator and marker of systemic inflammation, the authors suggested that their findings may support the hypothesis that inflammatory reaction can be the major factor in the occurrence and development of atherosclerosis. Likewise, Sisckovick et al. [41] revealed that the presence of IgG antibodies to HSV-1 was associated with two times increased risk of MI and CHD death. Notably, these results were achieved after adjustment for age, sex, cigarette smoking, diabetes, hypertension, HDL cholesterol, BMI, physical activity, and years of education. Rupprecht et al. [42] examined 78 German patients with CHD dead due to cardiovascular causes and 932 subjects with CHD survived for 4 years. The authors revealed that seropositivity to HSV-2 was significantly more frequent in subjects who dead in 4 following years, as compared to survivors. After adjustment for age, sex, smoking status, diabetes, HDL cholesterol, number of stenoses, invasive treatment, antihypertensive treatment and statin intake at enrollment, the risk for future cardiac death increased by two times. At the same time, no positive results were obtained for HSV-1 in this study. Interestingly, the authors indicated that HSV status was predictive of future cardiovascular events irrespective of the CRP level, which suggests that it may contribute to the atherosclerosis independent of an inflammatory response. The similar study was conducted by Zhu et al. [43], who followed up 890 mixed US patients suffering from CHD for 3 years. The authors demonstrated that seropositivity to either HSV-1 or HSV-2 was associated with 50 % increased risk of MI or death after adjustment for age, sex, number of vessels, presentation, diabetes, smoking, hyperlipidemia, hypertension, renal failure, and family history of CHD. Sun et al. [44] examined presence of HSV-2 among 488 Chinese patients with essential hypertension and among 756 normotensive subjects. After adjustment for age, male sex, smoking status, body mass index, dyslipidemia, diabetes and CHD, the authors revealed that carriers of HSV-2 have a 40 % increased risk for development of hypertension. Pesonen et al. [45] demonstrated an association between the presence of HSV and CHD risk among 335 Swedish subjects with unstable angina and MI. Elkind et al. [46] revealed that the presence of either HSV-1 or HSV-2 increases the risk of ischemic stroke by 35–60 % (1,625 mixed US subjects were followed-up for 7 years). Nikolopoulou et al. [47] reported that 150 Greek patients with CHD were more likely to have anti-HSV IgG in comparison with 49 healthy individuals.

Interestingly, series of studies provide evidence that precisely HSV-1 but not HSV-2 play an important role in the modulation of CVD risk. First, Kis et al. [48] demonstrated statistically significant associations between the presence of HSV-1, but not HSV-2, and almost fourfold increased ischemic stroke risk in 111 Hungarian subjects. Likewise, Jafarzadeh et al. [49] indicated a correlation between the presence of HSV-1, but not HSV-2, and threefold elevated risk of CHD and MI in 120 Iranian subjects as compared to 60 healthy controls. Similarly, Georges et al. [50] observed a 90 % increased risk of acute MI and angina among 991 German patients after adjustment for age, sex, body mass index, smoking status, diabetes, hypertension and high-density lipoprotein cholesterol level.

Finally, nine case-control studies obtained null results evaluating the impact of HSV on various cardiovascular events. Rothenbacher et al. [51] assessed the association between seropositivity to HSV and CHD in 312 patients with angiographically proven CHD and in 479 age and sex matched controls. The authors found statistically significant association between the seropositivity for HSV and CHD, but these associations became insignificant after adjustment for a variety of potential confounders. Similarly, no positive associations were observed in the findings by Heltai et al. [52], Sheehan et al. [53], Schlitt et al. [54], Lenzi et al. [55], Guan et al. [56, 57], Al-Ghamdi [58], Mundkur et al. [59], and Prasad et al. [60].

2.4 Discussion

More than 25 years have passed since HSV began to be studied as a potential etiologic agent of atherosclerosis. The intensive research in this field, specifically in recent years, is very remarkable. Multiple epidemiological, biological and clinical studies have been conducted in order to shed the light on this intriguing issue (Tables 2.1 and 2.2). Some investigators reveal positive associations while others do not. In this chapter we accumulate and generalize current data devoted to this topic. The majority of findings do not support the suggestion that HSV may contribute to the increased risk of atherosclerosis and related diseases. A lot of case-control studies demonstrated that the risk of CVDs was increased in HSV-positive individuals, but many investigations obtained null results as well. Moreover, in order to establish a positive correlation, viruses should be detected in the significant share of plaques from atherosclerotic species; however, they were actually found in single cases. It should be also noted that the results do not differ significantly in various ethnicities and populations. Therefore, despite there being some fundamental mechanisms that indicate HSV may lead to occurrence and progression of atherosclerosis (Fig. 2.1), current data are unable to demonstrate any reliable and firm connections. Unfortunately, the results achieved by the majority of studies are difficult to interpret due to revealing of only HSV-1 or HSV-2 in these investigations. Moreover, the authors sometimes did not specify which type of virus they determined in their study. Because of this, it is hard to identify which type of HSV may play a greater/less role in the etiology of atherosclerosis and related diseases, although series of studies indicate the major role of HSV-1. It is also important to note that currently there are no published meta-analyses devoted to this issue, and their appearance may help to clarify the issue.

There are two hypotheses explaining the presence of a virus in an atherosclerotic plaque. The first hypothesis suggests that one and only pathogen is involved in the development of atherosclerosis, such as *Helicobacter pylori* promotes gastritis and gastric ulcer. The second one states that several (more than one) bacterial or viral agents may collectively result in atherogenesis. According to the Tables 2.1 and 2.2 it is hardly possible that HSV may be the sole causative agent of atherosclerosis; therefore, the first hypothesis is not applicable for HSV. Hereby, the second hypothesis seems more reliable, as the studies by Watt et al. [27], Mundkur et al. [59], Georges et al. [50], Zhu et al. [61], and Prasad et al. [60] demonstrate that pathogen burden (≥5 pathogens, including HSV) is an independent risk factor for CVD risk.

Additionally, several important features of study organization can be proposed for the credible determination of the possible association of HSV with atherosclerosis and related diseases. First, case-control and cohort studies should be conducted in different populations and ethnicities. Second, the research devoted to the immune or atherogenic mechanisms of the viral impact on atherosclerosis development should be performed in order to clarify interactions between virus-related mechanisms and other mechanisms of atherosclerosis. Third, studies should investigate both HSV types 1 and 2, to see the potential differences in the effect of these pathogens on atherogenesis. Fourth, it is feasible to perform a meta-analysis of all of the above-mentioned studies in order to clarify common patterns of the impact of HSV. To conclude, we suggest there is no substantial basis to assume that HSV directly promote atherosclerosis and related diseases; nevertheless, as a part of pathogen burden, it's role may be significant. Perhaps, further studies will determine the exact mechanisms and associations between HSV and CVDs.

Table 2.1 Studies revealing HSV in plaques or serum of patients with early or advanced atherosclerosis

Author, reference	Population	Sample source	Detection method	Cases/controls	Virus prevalence or odds ratio
Benditt et al. [4]	Mixed (US)	Coronary artery	In situ DNA hybridization	160/–	7.5 % (12) for HSV[a]
Gyorkey et al. [18]	Mixed (US)	Proximal aorta	In situ DNA hybridization	60/–	17 % (10) for HSV
Ghosh et al. [20]	Mixed (US)	Coronary artery	In situ DNA hybridization	4/–	50 % (2) for HSV-2
Yamashiroya et al. [19]	Mixed (US)	Thoracic aorta, coronary artery, serum	In situ DNA hybridization, serologic methods	20/–	40 % (8) for HSV-2
Jackel et al. [21]	Causasian (German)	Endomyocardial biopsy	In situ DNA hybridization	29/–	25 % (7) for HSV
Sortie et al. [22]	Mixed (US)	Serum	Serologic methods	340/340	OR = 1.21 for HSV-1 ($P=0.45$) and 0.61 ($P=0.05$) for HSV-2
Raza-Ahmad et al. [23]	Caucasian (Canadian)	Coronary artery	In situ DNA hybridization	31/–	45 % (14) for HSV-2 and 3 % (1) for HSV-1
Chiu et al. [24]	Caucasian (Canadian)	Carotid artery, serum	Immunohistochemistry, serologic methods	76/–	10.5 % (8) for HSV-1
Espinola-Klein et al. [25]	Causasian (German)	Serum	Serologic methods	504/–	$P<0.0001$ for an association between HSV-2 and carotid IMT
Espinola-Klein et al. [26]	Causasian (German)	Serum	Serologic methods	572/–	$P<0.01$ for an association between HSV-2 and advanced atherosclerosis
Watt et al. [27]	Causasian (French)	Carotid artery	PCR	18/–	0 % for HSV-1 and 16 % (3) for HSV-2
Shi and Tokunaga [28]	Asian (Japanese)	Aorta	PCR, in situ DNA hybridization, Southern blotting	10/23	80 % (8)/13 % (3) for HSV-1

Study	Ethnicity	Tissue	Method	n (cases/controls)	Result
Kwon et al. [32]	Asian (Korean)	Carotid artery	PCR, rtPCR	128/20	0 % for HSV-1 and 0 % for HSV-2 in cases and controls
Skowasch et al. [33]	Causasian (German)	Carotid artery, serum	Immunohistochemistry, serologic methods	21/20	0 % for HSV in cases and controls
Ibrahim et al. [29]	Caucasian (Syrian)	Carotid artery, coronary artery	PCR	48/66	35 % (17)/9 % (6) $P=0.006$ for HSV-1
Muller et al. [30]	Caucasian (German)	Carotid artery, serum	PCR, serologic methods	109/–	97.1 % (105) in serum and 3.9 % (2) in plaques for HSV
Altun et al. [37]	Caucasian (Turkish)	Coronary artery	PCR	28/22	0 %/0 % for HSV-1 and HSV-2
Hagiwara et al. [35]	Asian (Japanese)	Carotid artery	PCR	50/–	4 % (2) for HSV
Reszka et al. [31]	Caucasian (Polish)	Aorta	PCR	40/20	67.5 % (27)/85 % (17) for HSV
Voorend et al. [34]	Caucasian (Dutch)	Basilar artery, middle cerebral artery	PCR	19/–	0 % for HSV-1
Szklo et al. [38]	Mixed (USA)	Serum	Serologic methods	1,056 patients w/o CVDs	84.9 % (847) for HSV, lack of association
Tremolada et al. [36]	Caucasian (Italian)	Carotid artery	PCR	17/–	0 % for HSV-1

[a]HSV type (1 or 2) has not been established if not specified

Table 2.2 Studies revealing possible associations between HSV and CVDs

Author, reference	Population	Condition or disease	Cases/controls	Virus prevalence in cases/controls and ORs with 95 % CIs
Ridker et al. [39]	Mixed (US)	MI and thromboembolic stroke	372 MI/271 stroke/643 controls	69.0 % (443)/69.4 % (446); OR: 0.92 (0.7–1.2) for MI and 1.1 (0.7–1.6) for stroke
Roivainen et al. [40]	Caucasian (Finnish)	MI and coronary death	241/241	97.9 % (236)/97.9 % (236); OR: 2.07 (1.20–3.56) for high HSV-1 levels adjusted for age and smoking status. OR: 25.44 (2.94–220.3) for high HSV-1 and CRP levels
Siscovick et al. [41]	Mixed (US)	MI and CHD death	231/405	89.6 % (207)/82.2 % (333); OR: 2.0 (1.1–3.6) adjusted for age, sex, and other covarities
Rupprecht et al. [42]	Caucasian (German)	CHD	78 CHD subjects dead from CVDs/932 CHD survivors	94.9 %/92.7 % for HSV-1 (P>0.005); 25.6 %/13.5 % for HSV-2; OR: 2.0 (1.01–4.0) adjusted for age, sex, and other covarities
Zhu et al. [43]	Mixed (US)	CHD, MI	167 CHD subjects undergone MI or death/723 CHD survivors	92.2 % (154)/84.5 % (610) for HSV-1; 80.2 % (134)/70.8 % (512); OR: 1.57 (0.88–2.80) for HSV1 and 1.51 (1.02–2.23) for HSV-2 for MI or death, OR: 1.57 (0.88–2.80) for HSV1 and 1.51 (1.02–2.23) for HSV-2 adjusted for age, sex, and other covarities
Prasad et al. [60]	Mixed (US)	CHD	N/A[a]	85.8 %/79 % for HSV-1 (P>0.05) lack of association
Georges et al. [50]	Caucasian (German)	Acute MI, angina	820 angina/171 AMI/333 controls	94 % (931)/84 % (279) for HSV-1 (P<0.05); OR: 1.9 (1.2–3.1) adjusted for age, sex, and other covarities 14 % (139)/8 % (27) for HSV-2 (P>0.05) lack of association
Rothenbacher et al. [51]	Caucasian (German)	CHD	312/479	21.8 % (68)/13.6 % (65) for HSV[b] (P=0.0003); lack of association after adjustment
Sun et al. [44]	Asian (Chinese)	Hypertension	488/756	38.3 % (187)/29.8 % (225) for HSV-2 (P=0.002); OR: 1.4 (1.1–1.8) adjusted for age, sex, and other covarities

Reference	Population	Disease	Cases/Controls	Results
Heltai et al. [52]	Caucasian (Hungarian)	Acute MI, stable effort angina	40 AMI/43 SEA/46 controls	Lack of association for HSV-1[a]
Sheehan et al. [53]	Caucasian (Irish)	Acute coronary syndrome	227/277	92.0 % (207)/90.3 % (243) for HSV ($P>0.05$) lack of association
Schlitt et al. [54]	Caucasian (German)	CHD	184/52	1.84 % (1)/0 % for HSV ($P>0.005$) lack of association
Lenzi et al. [55]	Caucasian (Italian)	CHD	80/160	96.2 % (77)/87.5(140 for HSV-1 ($P>0.05$) lack of association
Pesonen et al. [45]	Caucasian (Swedish)	MI, unstable angina	335/335	OR: 2.24 ($P<0.001$) for HSV
Kis et al. [48]	Caucasian (Hungarian)	Ischemic stroke	59/52	40.7 % (24)/15.7 % (8) for HSV-1 ($P=0.06$); OR: 3.68 (1.47–9.20) · 39.0 % (23)/27.4 % (14) for HSV-2 ($P>0.05$) lack of association
Nikolopoulou et al. [47]	Caucasian (Greek)	Acute MI, CHD	150 CHD/138 AMI/49 controls	97.2 % (146)/89.6 % (44) for HSV ($P<0.05$) in relation to CHD
Elkind et al. [46]	Mixed (US)	Ischemic stroke, MI, CHD death	67 stroke/98 MI/150 vascular death/705 survivors	86.3 % (1,402); OR: 1.35 (0.59–3.07) for HSV-1 adjusted for age, sex, and other covariates · 57.1 % (928); 1.59 (0.91–2.76) for HSV-2 adjusted for age, sex, and other covarities
Guan et al. [56]	Asian (Chinese)	Acute MI	N/A[a]	Lack of association for HSV-1 and HSV-2[a]
Jafarzadeh et al. [49]	Caucasian (Iranian)	IHD	120/60	60.8 % (73)/33.3 % (20) for HSV-1 ($P<0.005$); OR: 3.10 (1.62–5.95)
Guan et al. [57]	Asian (Chinese)	Acute MI	120/160	28.3 % (34)/28.3 % (17) for HSV-2 ($P=1.0$) Lack of association for HSV-1 and HSV-2[a]
Al-Ghamdi [58]	Caucasian (Saudi)	CHD, cerebral stroke, peripheral artery disease	40 CHD/20 stroke/15 PAD/15 controls	Lack of association for HSV-1[a]
Mundkur et al. [59]	Indian	CHD	433/433	55.5 (240)/49.5 % (214) for HSV ($P>0.05$) lack of association

[a]Data were not available

[b]HSV type (1 or 2) has not been established if not specified

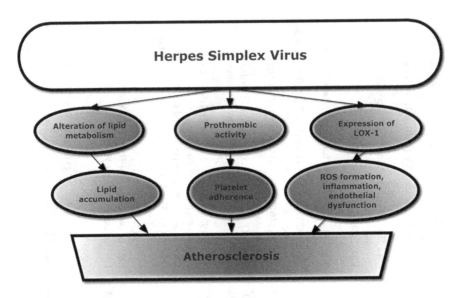

Fig. 2.1 Mechanisms of HSV-driven atherosclerosis

References

1. Kieff E, Hoyer B, Bachenheimer S, et al. Genetic relatedness of type 1 and type 2 herpes simplex viruses. J Virol. 1972;9(5):738–45.
2. Chayavichitsilp P, Buckwalter JV, Krakowski AC, et al. Herpes simplex. Pediatr Rev. 2009;30(4):119–29.
3. Blacklow NR, Rose FB, Whalen RA. Organ culture of human aorta: prolonged survival with support of viral replication. J Infect Dis. 1975;131(5):575–8.
4. Benditt EP, Barrett T, McDougall JK. Viruses in the etiology of atherosclerosis. Proc Natl Acad Sci U S A. 1983;80(20):6386–9.
5. Minick CR, Fabricant CG, Fabricant J, et al. Atheroarteriosclerosis induced by infection with a herpesvirus. Am J Pathol. 1979;96(3):673–706.
6. Hajjar DP, Fabricant CG, Minick CR, Fabricant J. Virus-induced atherosclerosis. Herpesvirus infection alters aortic cholesterol metabolism and accumulation. Am J Pathol. 1986;122(1):62–70.
7. Fabricant CG, Hajjar DP, Minick CR, et al. Marek's disease herpesvirus infection enhances cholesterol and cholesteryl ester accumulation in cultured avian arterial smooth muscle cells. Am J Pathol. 1981;105:176–84.
8. Hajjar DP, Falcone DJ, Fabricant CG, et al. Altered cholesteryl ester cycle is associated with lipid accumulation in herpesvirus-infected avian arterial smooth muscle cells. J Biol Chem. 1985;260:6124–8.
9. Hajjar DP, Pomerantz KB, Falcone DJ, et al. Herpes simplex virus infection in human arterial cells.Implications in arteriosclerosis. J Clin Invest. 1987;80(5):1317–21.
10. Visser MR, Tracy PB, Vercellotti GM, Goodman JL, et al. Enhanced thrombin generation and platelet binding on herpes simplex virus-infected endothelium. Proc Natl Acad Sci U S A. 1988;85(21):8227–30.
11. Key NS, Vercellotti GM, Winkelmann JC, Moldow CF, Goodman JL, Esmon NL, Esmon CT, Jacob HS. Infection of vascular endothelial cells with herpes simplex virus enhances tissue

factor activity and reduces thrombomodulin expression. Proc Natl Acad Sci U S A. 1990;87(18):7095–9.

12. Etingin OR, Silverstein RL, Hajjar DP. Identification of a monocyte receptor on herpesvirus-infected endothelial cells. Proc Natl Acad Sci U S A. 1991;88(16):7200–3.

13. Chirathaworn C, Pongpanich A, Poovorawan Y. Herpes simplex virus 1 induced LOX-1 expression in an endothelial cell line, ECV 304. Viral Immunol. 2004;17(2):308–14.

14. Morawietz H. LOX-1 and atherosclerosis: proof of concept in LOX-1-knockout mice. Circ Res. 2007;100(11):1534–6.

15. Morawietz H, Duerrschmidt N, Niemann B, Galle J, Sawamura T, Holtz J. Induction of the oxLDL receptor LOX-1 by endothelin-1 in human endothelial cells. Biochem Biophys Res Commun. 2001;284(4):961–5.

16. Mehta JL, Sanada N, Hu CP, Chen J, Dandapat A, Sugawara F, Satoh H, Inoue K, Kawase Y, Jishage K, Suzuki H, Takeya M, Schnackenberg L, Beger R, Hermonat PL, Thomas M, Sawamura T. Deletion of LOX1 reduces atherogenesis in LDLR knockout mice fed high cholesterol diet. Circ Res. 2007;100(11):1634–42.

17. Span AH, van Dam-Mieras MC, Mullers W, Endert J, Muller AD, Bruggeman CA. The effect of virus infection on the adherence of leukocytes or platelets to endothelial cells. Eur J Clin Invest. 1991;21(3):331–8.

18. Gyorkey F, Melnick JL, Guinn GA, Gyorkey P, DeBakey ME. Herpesviridae in the endothelial and smooth muscle cells of the proximal aorta in arteriosclerotic patients. Exp Mol Pathol. 1984;40(3):328–39.

19. Yamashiroya HM, Ghosh L, Yang R, Robertson Jr AL. Herpesviridae in the coronary arteries and aorta of young trauma victims. Am J Pathol. 1988;130(1):71–9.

20. Ghosh L, Yamashiroya H, Yang R, Garber S, Gabrovsek J, Robertson Jr AL. Herpes simplex virus antigen in human atheromatous lesions (Abstr). Fed Proc. 1985;44:1139.

21. Jäkel KT, Löning T, Arndt R, Rödiger W. Rejection, herpesvirus infection, and Ki67 expression in endomyocardial biopsy specimens from heart transplant recipients. Pathol Res Pract. 1992;188(1–2):27–36.

22. Sorlie PD, Adam E, Melnick SL, Folsom A, Skelton T, Chambless LE, Barnes R, Melnick JL. Cytomegalovirus/herpesvirus and carotid atherosclerosis: the ARIC Study. J Med Virol. 1994;42(1):33–7.

23. Raza-Ahmad A, Klassen GA, Murphy DA, Sullivan JA, Kinley CE, Landymore RW, Wood JR. Evidence of type 2 herpes simplex infection in human coronary arteries at the time of coronary artery bypass surgery. Can J Cardiol. 1995;11(11):1025–9.

24. Chiu B, Viira E, Tucker W, Fong IW. Chlamydia pneumoniae, cytomegalovirus, and herpes simplex virusin atherosclerosis of the carotid artery. Circulation. 1997;96(7):2144–8.

25. Espinola-Klein C, Rupprecht HJ, Blankenberg S, Bickel C, Kopp H, Rippin G, Hafner G, Pfeifer U, Meyer J. Are morphological or functional changes in the carotid artery wall associated with Chlamydia pneumoniae, Helicobacter pylori, cytomegalovirus, or herpes simplex virus infection? Stroke. 2000;31(9):2127–33.

26. Espinola-Klein C, Rupprecht HJ, Blankenberg S, Bickel C, Kopp H, Rippin G, Victor A, Hafner G, Schlumberger W, Meyer J, AtheroGene Investigators. Impact of infectious burden on extent and long-term prognosis of atherosclerosis. Circulation. 2002;105(1):15–21.

27. Watt S, Aesch B, Lanotte P, Tranquart F, Quentin R. Viral and bacterial DNA in carotid atherosclerotic lesions. Eur J Clin Microbiol Infect Dis. 2003;22(2):99–105.

28. Shi Y, Tokunaga O. Herpesvirus (HSV-1, EBV and CMV) infections in atherosclerotic compared with non-atherosclerotic aortic tissue. Pathol Int. 2002;52(1):31–9.

29. Ibrahim AI, Obeid MT, Jouma MJ, Moasis GA, Al-Richane WL, Kindermann I, Boehm M, Roemer K, Mueller-Lantzsch N, Gärtner BC. Detection of herpes simplex virus, cytomegalovirus and Epstein-Barr virus DNA in atherosclerotic plaques and inunaffected bypass grafts. J ClinVirol. 2005;32(1):29–32.

30. Müller BT, Huber R, Henrich B, Adams O, Berns G, Siebler M, Jander S, Müller W, Loncar R, Godehardt E, Sandmann W. Chlamydia pneumoniae, herpes simplex virus and

cytomegalovirus in symptomatic and asymptomatic high-grade internal carotid artery stenosis. Does infection influence plaque stability? Vasa. 2005;34(3):163–9.

31. Reszka E, Jegier B, Wasowicz W, Lelonek M, Banach M, Jaszewski R. Detection of infectious agents by polymerase chain reaction in human aortic wall. Cardiovasc Pathol. 2008;17(5):297–302.

32. Kwon TW, Kim DK, Ye JS, Lee WJ, Moon MS, Joo CH, Lee H, Kim YK. Detection of enterovirus, cytomegalovirus, and Chlamydia pneumonia in atheromas. J Microbiol. 2004;42(4):299–304.

33. Skowasch D, Jabs A, Andrié R, Dinkelbach S, Schiele TM, Wernert N, Lüderitz B, Bauriedel G. Pathogen burden, inflammation, proliferation and apoptosis in human instent restenosis. Tissue characteristicscompared to primary atherosclerosis. J Vasc Res. 2004;41(6):525–34.

34. Voorend M, van der Ven AJ, Kubat B, Lodder J, Bruggeman CA. Limited role for C. pneumoniae, CMV and HSV-1 in cerebral large and small vessel atherosclerosis. Open Neurol J. 2008;2:39–44.

35. Hagiwara N, Toyoda K, Inoue T, Shimada H, Ibayashi S, Iida M, Okada Y. Lack of association between infectious burden and carotid atherosclerosis in Japanese patients. J Stroke Cerebrovasc Dis. 2007;16(4):145–52.

36. Tremolada S, Delbue S, Ferraresso M, Carloni C, Elia F, Larocca S, Bortolani E, Ferrante P. Search for genomic sequences of microbial agents in atherosclerotic plaques. Int J Immunopathol Pharmacol. 2011;24(1):243–6.

37. Altun B, Rota S, Demircin M, Reşatoğlu A, Yener A, Bozdayi G. Investigation of herpes group and hepatitis A virus nucleic acids in the atherome plaque samples of patients withcoronary arterial disease. Mikrobiyol Bul. 2007;41(4):537–44.

38. Szklo M, Ding J, Tsai MY, Cushman M, Polak JF, Lima J, Barr RG, Sharrett AR. Individual pathogens, pathogen burden and markers of subclinical atherosclerosis: the Multi-Ethnic Study of Atherosclerosis. J Cardiovasc Med (Hagerstown). 2009;10(10):747–51.

39. Ridker PM, Hennekens CH, Stampfer MJ, Wang F. Prospective study of herpes simplex virus, cytomegalovirus, and the risk of future myocardial infarction and stroke. Circulation. 1998;98(25):2796–9.

40. Roivainen M, Viik-Kajander M, Palosuo T, Toivanen P, Leinonen M, Saikku P, Tenkanen L, Manninen V, Hovi T, Mänttäri M. Infections, inflammation, and the risk of coronary heart disease. Circulation. 2000;101(3):252–7.

41. Siscovick DS, Schwartz SM, Corey L, Grayston JT, Ashley R, Wang SP, Psaty BM, Tracy RP, Kuller LH, Kronmal RA. Chlamydia pneumoniae, herpes simplex virus type 1, and cytomegalovirus and incident myocardial infarction andcoronary heart disease death in older adults: the Cardiovascular Health Study. Circulation. 2000;102(19):2335–40.

42. Rupprecht HJ, Blankenberg S, Bickel C, Rippin G, Hafner G, Prellwitz W, Schlumberger W, Meyer J, AtheroGene Investigators. Impact of viral and bacterial infectious burden on long-term prognosis in patients with coronary artery disease. Circulation. 2001;104(1):25–31.

43. Zhu J, Nieto FJ, Horne BD, Anderson JL, Muhlestein JB, Epstein SE. Prospective study of pathogen burden and risk of myocardial infarction or death. Circulation. 2001;103(1):45–51.

44. Sun Y, Pei W, Wu Y, Jing Z, Zhang J, Wang G. Herpes simplex virus type 2 infection is a risk factor for hypertension. Hypertens Res. 2004;27(8):541–4.

45. Pesonen E, Andsberg E, Ohlin H, Puolakkainen M, Rautelin H, Sarna S, Persson K. Dual role of infections as risk factors for coronary heart disease. Atherosclerosis. 2007;192(2):370–5.

46. Elkind MS, Ramakrishnan P, Moon YP, Boden-Albala B, Liu KM, Spitalnik SL, Rundek T, Sacco RL, Paik MC. Infectious burden and risk of stroke: the Northern Manhattan Study. Arch Neurol. 2010;67(1):33–8.

47. Nikolopoulou A, Tousoulis D, Antoniades C, Petroheilou K, Vasiliadou C, Papageorgiou N, Koniari K, Stefanadi E, Latsios G, Siasos G, Stefanadis C. Common community infections and the risk for coronary artery disease and acute myocardial infarction: evidence for chronic over-expression of tumor necrosis factor alpha and vascular cells adhesion molecule-1. Int J Cardiol. 2008;130(2):246–50.

48. Kis Z, Sas K, Gyulai Z, Treso B, Petrovay F, Kapusinszky B, Csire M, Endresz V, Burian K, Mandi Y, Vecsei L, Gonczol E. Chronic infections and genetic factors in the development of ischemic stroke. New Microbiol. 2007;30(3):213–20.

49. Jafarzadeh A, Nemati M, Tahmasbi M, Ahmadi P, Rezayati MT, Sayadi AR. The association between infection burden in Iranian patients with acute myocardial infarction and unstablean-gina. Acta Med Indones. 2011;43(2):105–11.

50. Georges JL, Rupprecht HJ, Blankenberg S, Poirier O, Bickel C, Hafner G, Nicaud V, Meyer J, Cambien F, Tiret L, AtheroGene Group. Impact of pathogen burden in patients with coronary artery disease in relation to systemic inflammation andvariation in genes encoding cytokines. Am J Cardiol. 2003;92(5):515–21.

51. Rothenbacher D, Brenner H, Hoffmeister A, Mertens T, Persson K, Koenig W. Relationship between infectious burden, systemic inflammatory response, and risk of stable coronary arterydisease: role of confounding and reference group. Atherosclerosis. 2003;170(2):339–45.

52. Heltai K, Kis Z, Burian K, Endresz V, Veres A, Ludwig E, Gönczöl E, Valyi-Nagy I. Elevated antibody levels against Chlamydia pneumoniae, human HSP60 and mycobacterial HSP65 are independent risk factors in myocardial infarction and ischaemic heart disease. Atherosclerosis. 2004;173(2):339–46.

53. Sheehan J, Kearney PM, Sullivan SO, Mongan C, Kelly E, Perry IJ. Acute coronary syndrome and chronic infection in the Cork coronary care case-control study. Heart. 2005;91(1):19–22.

54. Schlitt A, Blankenberg S, Weise K, Gärtner BC, Mehrer T, Peetz D, Meyer J, Darius H, Rupprecht HJ. Herpes virus DNA (Epstein-Barr virus, herpes simplex virus, cytomegalovirus) in circulating monocytes of patients with coronary artery disease. Acta Cardiol. 2005;60(6):605–10.

55. Lenzi C, Palazzuoli A, Giordano N, Alegente G, Gonnelli C, Campagna MS, Santucci A, Sozzi M, Papakostas P, Rollo F, Nuti R, Figura N. H pylori infection and systemic antibodies to CagA and heat shock protein 60 in patients with coronary heart disease. World J Gastroenterol. 2006;12(48):7815–20.

56. Guan XR, Jiang LX, Ma XH, Wang LF, Quan H, Li HY. Respiratory syncytial virus infection and risk of acute myocardial infarction. Am J Med Sci. 2010;340(5):356–9.

57. Guan X, Yang W, Sun X, Wang L, Ma B, Li H, Zhou J. Association of influenza virus infection and inflammatory cytokines with acute myocardial infarction. Inflamm Res. 2012;61(6):591–8.

58. Al-Ghamdi A. Role of herpes simplex virus-1, cytomegalovirus and Epstein-Barr virus in atherosclerosis. Pak J Pharm Sci. 2012;25(1):89–97.

59. Mundkur LA, Rao VS, Hebbagudi S, Shanker J, Shivanandan H, Nagaraj RK, Kakkar VV. Pathogen burden, cytomegalovirus infection and inflammatory markers in the risk of prema-ture coronary artery disease in individuals of Indian origin. Exp Clin Cardiol. 2012;17(2):63–8.

60. Prasad A, Zhu J, Halcox JP, Waclawiw MA, Epstein SE, Quyyumi AA. Predisposition to ath-erosclerosis by infections: role of endothelial dysfunction. Circulation. 2002;106(2):184–90.

61. Zhu J, Quyyumi AA, Norman JE, Csako G, Waclawiw MA, Shearer GM, Epstein SE. Effects of total pathogen burden on coronary artery disease risk and C reactive protein levels. Am J Cardiol. 2000;85(2):140–6.

Chapter 3
The Role of Epstein-Barr Virus in Atherosclerosis and Related Diseases

Abstract A hypothesis of the possible role of Epstein-Barr virus (EBV) in the etiology of atherosclerosis has been proposed almost 30 years ago. The possible mechanisms of this connection can include induction and/or promotion of the proliferation of artery wall intimal smooth muscle cells by inflammatory injury or induction of genomic alterations leading to clonal expansion of intimal smooth muscle cell populations. In this review, we analyzed the feasible association between EBV and atherosclerosis. The results of certain investigators support the point of view that EBV-infection may influence the development of atherosclerosis; however, the majority of studies disprove this hypothesis. Almost all the studies where the positive association was found should not be established as entirely credible. Moreover, contradicting results were observed even in the same populations that refutes the hypothesis about the population dependence on the association between EBV and atherosclerosis. On the whole, the situation with the feasible association between EBV and atherosclerosis seems to be following: albeit there is a number of basic mechanisms supporting this hypothesis, current epidemiological data do not evidence in favor of this association. At the moment, it is not possible to establish EBV as a probable causative agent of atherosclerosis until reliable epidemiological studies devoted to the problem will confirm this hypothesis.

3.1 Epstein-Barr Virus and Atherosclerosis: Published Data from Basic and Epidemiological Investigations

Epstein-Barr virus (EBV, human herpesvirus-4, HHV-4) is a member of the γ-herpesvirus family, identified by Epstein and Barr in 1964 [1]. Subsequent studies showed that it is the causative agent of infectious mononucleosis, and over 90 % of the worldwide population is asymptomatically infected by EBV [2, 3]. EBV infects both B cells and epithelial cells, and intermittent reactivation and virus replication in the oro- and nasopharyngeal epithelial cells provide the spreading of EBV infection and latent infection in B lymphocytes [2, 3]. The infection is usually self-limited and controlled by the strongly elevated T cell immune response [4]. If the infection occurs in adolescence or adulthood, up to 50 % T cells in the host can be specific to the virus, which may cause the clinical symptom of infectious mononucleosis [4]. EBV then persists latently in the host within long-life memory B cells [4]. EBV has also been demonstrated in T and NK cells proliferating monoclonally or oligoclonally in a group of diseases including chronic active EBV infection and EBV-associated hemophagocytic lymphohistiocytosis [5–7]. In addition, EBV may infect endothelial cells [8], fibroblasts [9] and dendritic cells as well [10].

The underlying mechanism of the chronic inflammatory process in atherosclerosis is still unknown in a significant extent. As a possible trigger, several studies have suggested that various viruses and bacteria are associated with atherosclerotic diseases. Benditt et al. [11] were one of the very first who searched the possible role of viruses in the etiology of human atherosclerosis. In their pioneer study, they investigated the presence of viral genomes in arterial tissues by the method of *in situ* hybridization. Based on the fact that chickens infected with Marek disease virus, a herpesvirus, suffered from atherosclerotic lesions after infection, they focused on the possible presence of herpesviruses in human artery wall tissue, particularly, in atherosclerotic samples. The authors used herpesvirus probes on aortic wall specimens removed from patients undergoing coronary bypass surgery. They found herpes simplex viral mRNA in 13 specimens. Importantly, in certain cases the samples represented early stages of atherogenesis. In addition, the authors have also demonstrated that herpes simplex virus can infect human fetal smooth muscle cells in culture. However, Epstein-Barr virus (EBV) genome has not been observed in any of the specimens. In this paper, the authors proposed a hypothesis that herpesviruses (and possibly EBV) can induce and/or promote proliferation of artery wall intimal smooth muscle cells by inflammatory injury or cause genomic alterations leading to clonal expansion of intimal smooth muscle cell populations. According to their point of view, a hypothesis of the viral etiology of atherosclerosis may explain several features in the occurrence of atherosclerosis and thrombosis: intimal cell proliferation in the absence of certain common risk factors, the clonal nature of cell populations observed in many human atherosclerotic lesions, a large genetic determination of the atherosclerosis risk, and the role of certain environmental factors in atherosclerosis development. Likewise, Yamashiroya et al. [12] did not detect EBV nucleic acids and capsid antigen in the arteries of 20 young trauma victims with frequent atheromatous lesions using in situ DNA hybridization and ABC immunoperoxidase methods, respectively.

However, the hypothesis about the role of one of the most important herpesviruses, EBV, in the development of atherosclerosis was still alive, and some research groups have made an effort to find the association between EBV and atherosclerosis. In 1990, Musiani et al. [13] evaluated the immune response against EBV-induced antigens in the sera of 36 Italian patients with atherosclerosis and 36 matched controls, revealing significantly higher titres of EBV-specific antibodies and a significant increase in EBV-reactivated infections in the case group. So, they were the first who detected a positive connection between EBV and atherosclerosis. Nevertheless, the sample size was relatively small, and the prevalence of EBV-infection between cases and controls did not vary significantly. Three years later, Straka et al. [14] assessed coronary artery lesions in 43 Czech patients surviving for more than 3 months after heart transplantation. The authors detected lesions in 40 % of subjects (17 out of 43) and revealed an association with previous EBV-infection. Nonetheless, there was no control group in this study, that limits its reliability. In the same year, Delneste et al. [15] demonstrated the possibility of production of anti-endothelial cell antibodies by co-culture of EBV-infected human B cells with endothelial cells. Nevertheless, the subsequent study by Tanaka et al. [16] did not detect viral DNA and RNA in 37 atherosclerotic aneurysms and 16 normal aortic samples.

For the long 5 years there were no investigations in the discussing field, and possibly, it can be explained by the implausibility of the hypothesis. Only in 1999, EBV-containing T cells have been unexpectedly demonstrated in atherosclerotic abdominal aneurysms by Luo et al. [17]. Later, Horvath et al. [18] tested a possible role of EBV in the development of atherosclerosis in 244 Czech patients with ischemic heart disease (IHD) and 87 non-ischemic controls. In this investigation, DNA of EBV was detected in 59 % and 50 % of arterial walls of cases and controls, respectively. In addition, viral DNA was absent in venous samples. However, the differences between cases and controls were not statistically significant. In the prospective study, conducted by Rupprecht et al. [19] in German population, the influence of prior infection to EBV on the course of atherosclerosis was found (measured in IgA antibodies), showing the strongest association of all single pathogens with future events [(27.8 % in 78 patients who dead from cardiovascular causes and only 13.8 % in 932 survivors, mean follow-up 3.1 years, HR=2.8 (1.5–5.0) adjusted $P=0.001$)]. The authors suggested that infection-induced autoimmune responses may contribute to the development of atherosclerosis, and EBV may be a candidate trigger, since virus-induced antigens are part of B-cell membranes and thence may stimulate activity of T-cells. The similar results have been obtained in another prospective study by Espinola-Klein et al. [20], who studied 504 German patients with carotid atherosclerosis (mean follow-up 2.5 years) measuring intima-media thickness and prevalence of carotid artery stenosis. In this investigation, a positive correlation between anti-EBV IgA seropositivity and progression of atherosclerosis was observed [(12.6 % in 311 patients with no progression, 18.9 % in 116 individuals with progression of atherosclerosis, adjusted OR=1.40 (1.08–2.68), $P=0.01$]. However, in another prospective study by the same authors [21], based on 572 German patients with atherosclerosis (mean follow-up 3.2 years), they failed to replicate their previous results [adjusted OR of advanced atherosclerosis =0.89 (0.73–1.30), $P=0.21$].

Studies demonstrating a positive association between EBV and atherosclerosis stimulated an emergence of numerous investigations in the field by various research groups in different countries. Shi and Tokunaga [22] investigated the presence of viral DNA in aortic tissues from 10 atherosclerotic and 23 non-atherosclerotic aortic tissues collected from Japanese patients. In the case group, 8 out of 10 tissue samples (80 %) were EBV-positive, compared to only 3 out of 23 (13 %) in the control group. Authors detected viral DNA in cells of the upper part of the non-atherosclerotic aortic wall, and EBV nucleic acids were detected more extensively in atherosclerotic lesions in comparison with non-atherosclerotic tissue. Although the results were positive, the sample size was insufficient; therefore, this study could not be established as a substantive evidence of the association between EBV and atherosclerosis. In contrast, De Backer et al. [23] found no difference in prevalence of serum antibodies to EBV in 446 Belgian patients with coronary heart disease and in 892 age-, educational level- and industry-matched controls. In addition, Khairy et al. [24] did not find an association between chronic EBV-infection (measured by anti-EBV nuclear antigen) and endothelial dysfunction (measured by flow-mediated brachial vasodilation) in 65 Canadian male subjects with no risk factors or known coronary artery disease. Moreover, no dose-response trends between serum titers and endothelial function were found. Similar to these investigations, Pitiriga et al. [25] in Greece revealed no difference in EBV seropositivity and elevated blood pressure in 146 sustained hypertensives defined by 24 h ambulatory blood pressure monitoring and 54 normotensives. In their second study [26], they found no association of EBV seropositivity and carotid intima-media thickness in 340 subjects underwent 24-h ambulatory blood pressure (BP) monitoring, clinic BP measurements, and ultrasound carotid measurements. Andrie et al. [27] evaluated the intimal presence of EBV in human coronary atheroma, and they also assessed the effect of EBV-infection on the expression of hHSP60, one of the key proteins in immune-mediated pathogenesis of atherosclerosis. The authors collected 53 lesions in German patients with acute coronary syndrome and stable angina, revealing the presence of EBV proteins in 42 %. In addition, the presence of EBV was associated with the expression of hHSP60. However, the control group was absent in this study.

In the following years, a number of research groups did not reveal an association between EBV and atherosclerosis. For instance, Kwon et al. [28] did not find EBV DNA in atherosclerotic plaques from 128 Korean patients with atherosclerosis and from 20 arterial wall samples from the control group. Ibrahim et al. [29] in Syria assessed the presence of EBV DNA in 48 atherosclerotic plaques and in non-atherosclerotic vessels from the same patient (23 internal mammary arteries, 43 saphenous veins), founding EBV DNA only in 2 % of plaques. In the study by Lobzin et al. [30], who recruited 64 Russian patients with coronary artery disease and 38 healthy controls, diagnostically significant elevation of the serum levels of EBV-specific IgG antibodies was not associated with coronary artery disease progression. Schlitt et al. [31] found EBV DNA only in 9 out of 184 German patients with coronary artery disease and in 2 out of 52 healthy controls, and this difference was not significant. According to the study carried out in Italy by Lenzi et al. [32], there is no difference between EBV prevalence (measured by ELISA, IgG anti-EBV antibodies) in patients suffering from coronary artery disease (80 subjects) and healthy controls (160 subjects, in both cases the share of EBV-positive individuals was 92.5 %). Altun et al. [33] did not detect EBV DNA in atherome plaque samples from 28 Turkish patients with atherosclerotic heart disease but observed it in 2 out of 22 vein samples from the controls who had other vascular diseases. Bayram et al. [34] did not find EBV nucleic acids in 30 Turkish atherosclerotic and 30 non-atherosclerotic vascular samples from patients with coronary artery disease. Tremolada et al. [35] detected EBV DNA only in 3 out of 17 atherosclerotic plaques from carotid arteries of 17 patients, and Valmary et al. [36] in France were unsuccessful to detect the presence of EBV nucleic acids and proteins in 34 lungs explanted from 19 patients with end-stage pulmonary arterial hypertension.

However, certain basic studies sustained the hypothesis about the possible role of EBV in atherosclerosis. In the investigation by de Boer et al. [37], viral DNA was found in 15 out of 19 tissue samples (79 %). In addition, they revealed that EBV-specific cytotoxic T-cells can also be frequently observed in human atherosclerotic plaques (11 out of 19 samples, 58 %) and suggested that a T-cell response against EBV may play a role in the plaque inflammation. The authors demonstrated that EBV-specific T-cells were generated at the site of the plaque, affirming that a local EBV-specific T-cell response can contribute to the inflammatory process presumably related to the EBV-infected macrophages. This study provided a probable basis for the EBV-induced initiation of pro-atherogenic inflammatory response. It is also known that EBV infection may lead to the induction of NF-κB signaling pathway due to the activity of the membrane protein LMP-1 [38]. In addition, EBV can stimulate macrophages to produce the macrophage inflammatory protein-1, attracting B and T lymphocytes to the site of inflammation [39]. The research group of Waldman et al. [40] found that EBV-encoded dUTPase stimulated monocytes/macrophages to upregulate the expression of TNF-α through a NF-κB mechanism, promoting the upregulation of the surface expression of VCAM-1 and ICAM-1 (inflammation-associated endothelial adhesion molecules) and an upregulation of IL-6. According to these data, EBV may induce an inflammatory cascade. The significant association of BDI depression scores with antibody titers to the EBV-encoded dUTPase together with the significant relationship between plasma IL-6 levels and neutralizing antibody titers to the EBV-encoded dUTPase supports a connection of depressive symptoms, EBV reactivation, and the upregulation of proinflammatory cytokine synthesis such as IL-6, a well-known atherosclerosis risk factor [41]. Kempe et al. [42] assessed the expression of Epstein-Barr virus-induced gene 3 (Ebi3) in human atheromatous lesions and evaluated its transcriptional regulation in vascular cells. Smooth muscle cells of human endarterectomy specimens expressed Ebi3 as well as the IL-27alpha/p28 and IL-12alpha/p35 subunits. Primary aortic smooth muscle cells overexpressed Ebi3 in response to proinflammatory stimuli such as TNF-α and IFN-γ. The authors proposed a possible role of the Ebi3 in the atherogenesis either as homodimer or as IL-27/IL-35 heterodimer, suggesting that it may be an interesting therapeutic target for atherosclerosis treatment (IL-27 and IL-35 regulate T cells proliferation and differentiation, so they can be important regulators of atherosclerotic plaque formation as well). Apostolou et al. [43] evaluated the effects of EBV infection on lipid and lipoprotein pattern in 29 patients with infectious mononucleosis and in 30 controls. The level of total cholesterol, high-density lipoprotein cholesterol, low-density lipoprotein cholesterol, apolipoproteins (apo) A-I, B, C-III, and lipoprotein (a) was lower at baseline, whilst apoB/apoA-I ratio, triglyceride levels and cholesteryl-ester transfer protein activity were elevated in comparison with 4 months later. At baseline, higher levels of cytokines and the cholesterol concentration of small-dense low-density lipoprotein particles were observed, whereas low-density lipoprotein particle size was lower compared with follow-up. Activities of lipoprotein-associated phospholipase A2 and PON1 were similar at baseline and 4 months later. Four months after the resolution of myocardial infarction levels of triglycerides, apoE, apoC-III, lipoprotein (a), cholesterol concentration of small-dense low-density lipoprotein particles and cytokines as well as low-density lipoprotein particle size, apoB/apoA-I ratio, cholesteryl-ester transfer protein and Lp-PLA2 activities were similar to controls. PON1 activities both at baseline and 4 months later were lower in patients compared with controls. So, authors found that infectious mononucleosis is associated with atherogenic changes of lipids and lipoproteins that can be partially restored 4 months after its resolution.

In addition, Andrie et al. [44] found the presence of EBV by immunohistochemistry in 40 % of coronary atherectomy specimens from 60 primary lesions of 35 patients with acute coronary syndrome and 25 patients with stable angina and evaluated the presence of EBV by immunohistochemistry. In addition, they revealed in atherosclerotic lesions an increased expression of C-reactive protein, tissue factor, and hHSP60, which was associated with the number of infectious pathogen. However, this study was carried out without the group of control. Al-Ghamdi [45] assessed the levels of anti-EBV IgG antibodies among 75 patients with atherosclerotic vascular diseases and 15 healthy controls. He also evaluated the association of the level of these antibodies with features of lipid profile and level of high-sensitive C-reactive protein (hsCRP) in these patients. There were no significant differences regarding the presence of EBV-specific IgG antibodies among cases and controls, but the level of these antibodies was significantly elevated among patients with atherosclerotic vascular diseases. However, there was no significant correlation between the level of virus-specific IgG antibodies and lipid profile or hsCRP. All studies devoted to the problem of the association between EBV and atherosclerosis are summarized in Table 3.1.

3.2 Discussion

The question whether EBV is a possible cause of atherosclerosis remains obscure. As it can be seen from Table 3.1, the results of certain investigators support the point of view that EBV-infection may influence the development of atherosclerosis, but the majority of studies disprove this hypothesis. In this section, we will try to solve this discrepancy. First, it is important to note that the sample size in most of the investigations was relatively small, generally not exceeding 200 participants, and this limits the statistical power of these studies. Possibly, if the samples would be larger, the results could be different, and at least a few large prospective or retrospective studies are necessary for the objective view on the problem. Second, the methods used in different investigations vary significantly, from genomic (usually PCR, real-time PCR, or in situ hybridization) to proteomic (usually ELISA or microimmunofluorescence). Even within proteomic methods, various antibodies can be determined, either IgG or IgA (and IgM in rare cases), and it also affects the study results. For instance, the prevalence of EBV-infection according to the

Table 3.1 The association of EBV with atherosclerosis, coronary artery disease, acute coronary syndrome, angina, myocardial infarction, arterial hypertension, and stroke

Authors, reference, population	Method of EBV detection, sample source	Number of cases and controls	EBV prevalence in cases and controls	Association of EBV with atherosclerosis
Benditt et al. [11] US population	In situ hybridization, arterial specimens		0 %	No association
Yamashiroya et al. [12] US population	In situ DNA hybridization, ABC immunoperoxidase methods, arterial specimens	20 cases (atheromatous lesions)	0 %	No association
Musiani et al. [13] Italian population	Proteomic, serum	36 cases (atherosclerosis), 36 controls		Positive association (significantly higher titres of antibody against EBV-induced antigens and a significant increase in EBV-reactivated infections were found in atherosclerotic patients)
Straka et al. [14] Czech population		43 cases (coronary artery lesions)	40 %	Positive association
Tanaka et al. [16] Japanese population	PCR, arterial specimens	37 cases (atherosclerotic aortic aneurysms) and 16 controls	0 % in both cases and controls	No association
Horvath et al. [18] Czech population	PCR, IgM and IgG detection, arterial specimens	244 cases (ischemic heart disease), 87 controls	59 % in cases, 50 % in controls	No association
Rupprecht et al. [19] German population	IgG, IgA ELISA, serum	78 deaths (cardiovascular events), 932 controls (survivors), all they suffered from coronary heart disease	27.8 % in cases, 13.8 % in controls, HR=2.8 (1.5–5.0) adjusted $P=0.001$ (IgA antibodies) 100 % in cases, 98.7 % in controls (IgG antibodies)	Positive association

(continued)

Table 3.1 (continued)

Authors, reference, population	Method of EBV detection, sample source	Number of cases and controls	EBV prevalence in cases and controls	Association of EBV with atherosclerosis
Espinola-Klein et al. [20] German population	IgG, IgA ELISA, serum	116 cases (progression of carotid atherosclerosis), 311 controls (no progression of carotid atherosclerosis)	18.9 % in cases, 12.6 % in controls, adjusted OR=1.40 (1.08–2.68), P=0.01 (IgA antibodies) 99.1 % in cases, 97.7 % in controls (IgG antibodies)	Positive association
Espinola-Klein et al. [21] German population	IgG, IgA ELISA, serum	572 patients with atherosclerosis	Adjusted OR of advanced atherosclerosis =0.89 (0.73–1.30), P=0.21 (IgA antibodies)	No association
Shi and Tokunaga [22] Japanese population	PCR, southern blotting, in situ hybridization, arterial specimens	10 cases (atherosclerotic aortic tissues), 23 controls (non-atherosclerotic aortic tissues)	80 % in cases, 13 % in controls	Positive association
De Baker et al. [23] Belgian population	Proteomic, serum	446 cases (coronary heart disease), 892 controls		No association
Khairy et al. [24] Canadian population	Immunofluorescence (IgG), serum	65 patients	88.9 %	No association
Andrie et al. [27] German population	Immunohistochemistry, arterial specimens	33 cases with acute coronary syndrome, 20 cases with stable angina	42 %	No association
Pitiriga et al. [25] Greek population	ELISA, microimmunofluorescence, serum	146 cases (arterial hypertension), 54 controls		No association
Pitiriga et al. [26] Greek population	ELISA, microimmunofluorescence, serum	340 participants (arterial hypertension, normotension)		No association with intima-media thickness
Kwon et al. [28] Korean population	PCR, arterial specimens	128 cases (atherosclerosis), 20 controls	0 % in both cases and controls	No association

Study	Method, specimen	Cases/controls	Percentage	Association
Ibrahim et al. [29] Syrian population	Real-time PCR, vascular specimens	48 case atherosclerotic plaques and control non-atherosclerotic vessels from the same patient	2 % in cases, $P=1.0$	No association
Lobzin et al. [30] Russian population	IgG antibodies determination, serum	64 cases (coronary artery disease), 38 controls		No association
Schlitt et al. [31] German population	PCR, arterial specimens	74 cases with stable angina, 51 cases with unstable angina, 59 cases with myocardial infarction, 52 controls	4.9 % in cases, 3.8 % in controls, $P=0.752$	No association
Lenzi et al. [32] Italian population	IgG ELISA, serum	80 cases (coronary artery disease), 160 controls	92.5 % in both cases and controls	No association
De Boer et al. [37]	PCR, arterial specimens	19 cases (carotid atherosclerosis)	79 %	
Altun et al. [33] Turkish population	PCR, arterial specimens	28 cases (coronary artery disease), 22 controls (vascular diseases other than atherosclerosis)	0 % in cases, 9 % in controls	No association
Andrie et al. [44] German population	Immunohistochemistry, arterial specimens	35 cases (acute coronary syndrome), 25 cases (stable angina)	40 %	
Bayram et al. [34] Turkish population	PCR, arterial specimens	30 cases (coronary artery disease), 30 controls	0 %	No association
Tremolada et al. [35]	Quantitative real-time PCR, arterial specimens	17 cases (carotid atherosclerosis)	17.6 %	
Valmary et al. [36] French population	Immunohistochemistry, in situ hybridization, arterial specimens	19 cases (end-stage pulmonary arterial hypertension)	0 %	No association
Al-Ghamdi et al. [45] Saudi Arabian population	IgG by ELISA, serum	20 cases (acute coronary syndrome, 20 cases (coronary artery disease) 20 cases (cerebral stroke), 15 cases (peripheral arterial disease), 15 controls		Positive association (only the levels of EBV-specific IgG antibodies were significantly ($P<0.05$) elevated among patients with atherosclerosis)

detection of anti-EBV IgG antibodies varied from 92.5 to 100 % in cases and from 92.5 to 97.7 % in controls, whilst according to the detection of anti-EBV IgA antibodies it varied from 18.9 to 27.8 % in cases and from 12.6 to 13.8 % in controls. According to the PCR detection of EBV nucleic acids, the prevalence of EBV-infection varied from 0 to 80 % in cases and from 0 to 50 % in controls. The positive association of EBV with atherosclerosis was demonstrated in six studies with clearly defined case and control groups. However, in four of them only proteomic methods (detection of IgG or IgA anti-EBV antibodies in serum by ELISA or microimmunofluorescence) were used, and the only sample source was serum. In two out of these four studies differences in prevalence of IgG anti-EBV antibodies between cases and controls were not statistically significant (only the titres of IgG antibodies were higher in the case group), and in other two studies the differences between prevalence of anti-EBV IgG antibodies were also not significant (only differences in prevalence of IgA antibodies were significant; nevertheless, it just shows the insufficient sensitivity of the IgA antibodies). In one study, sample source and methods were not defined (article in Czech), and only in one study the positive association was revealed by application of genomic methods (PCR, in situ hybridization) using arterial specimens as a sample source. However, the sample size in this study was very low (10 cases and 23 controls), and it limits its robustness in a significant extent. So, all six studies with clearly defined case and control groups where the positive association between EBV and atherosclerosis was found raise questions in their reliability. Studies without control group can not be represented as an evidentiary standard, even if a possible positive association was detected. So, almost all the studies where the positive association was found should not be established as entirely credible. Third, it is possible that differences in populations may also affect an impact of EBV on the development of atherosclerosis. However, there is no consistent pattern in the distribution of positive results between distinct continents (4 in Central Europe, 1 in Japan, and 1 in Saudi Arabia), and negative results were also widely distributed among continents (3 in North America, 11 in Europe, 2 in Korea and Japan, and 1 in Syria). Even in the same country contrary results have been obtained (Germany, Italy, Japan) that refutes the hypothesis about the population dependence on the association between EBV and atherosclerosis.

On the whole, the situation with the feasible association between EBV and atherosclerosis seems to be following: albeit there is a number of basic mechanisms supporting this hypothesis, the data from numerous epidemiological studies do not evidence in favor of this association. It is also obvious that it is not possible to establish EBV as a probable causative agent of atherosclerosis until reliable epidemiological studies devoted to the problem will confirm this theory. Additional fundamental investigations analyzing existing and novel mechanisms of EBV-related atherosclerotic plaque formation are also desirable.

At the moment, it is unlikely that EBV may be an established cause of atherosclerosis, but the ground for speculations on this question still exists. Only future fundamental and epidemiological investigations may give an eventual answer and solve the problem. However, if EBV is a real cause of atherosclerosis, potential anti-EBV vaccines may reduce atherosclerosis incidence in the following decades.

References

1. Epstein MA, Achong BG, Barr YM. Virus particles in cultured lymphoblasts from Burkitt's lymphoma. Lancet. 1964;1:702–3.
2. Middeldorp JM, Brink AA, Van Den Brule AJ, Meijer CJ. Pathogenic roles for Epstein-Barr virus (EBV) gene products in EBV-associated proliferative disorders. Crit Rev Oncol Hematol. 2003;45:1–36.
3. Thompson MP, Kurzrock R. Epstein-Barr virus and cancer. Clin Cancer Res. 2004;10: 803–21.
4. Young LS, Rickinson AB. Epstein-Barr virus: 40 years on. Nat Rev Cancer. 2004;4:757–68.
5. Kikuta H, Taguchi Y, Tomizawa K, Kojima K, Kawamura N, Ishizaka A, Sakiyama Y, Matsumoto S, Imai S, Kinoshita T. Epstein-Barr virus genome-positive T lymphocytes in a boy with chronic active EBV infection associated with Kawasaki-like disease. Nature. 1988;333:455–7.
6. Jones JF, Shurin S, Abramowsky C, Tubbs RR, Sciotto CG, Wahl R, Sands J, Gottman D, Katz BZ, Sklar J. T-cell lymphomas containing Epstein-Barr viral DNA in patients with chronic Epstein-Barr virus infections. N Engl J Med. 1988;318:733–41.
7. Kawaguchi H, Miyashita T, Herbst H, Niedobitek G, Asada M, Tsuchida M, Hanada R, Kinoshita A, Sakurai M, Kobayashi N. Epstein-Barr virus-infected T lymphocytes in Epstein-Barr virus-associated hemophagocytic syndrome. J Clin Invest. 1993;92:1444–50.
8. Jones K, Rivera C, Sgadari C, Franklin J, Max EE, Bhatia K, Tosato G. Infection of human endothelial cells with Epstein-Barr virus. J Exp Med. 1995;182:1213–21.
9. Koide J, Takada K, Sugiura M, Sekine H, Ito T, Saito K, Mori S, Takeuchi T, Uchida S, Abe T. Spontaneous establishment of an Epstein-Barr virus-infected fibroblast line from the synovial tissue of a rheumatoid arthritis patient. J Virol. 1997;71:2478–81.
10. Severa M, Giacomini E, Gafa V, Anastasiadou E, Rizzo F, Corazzari M, Romagnoli A, Trivedi P, Fimia GM, Coccia EM. EBV stimulates TLR- and autophagy-dependent pathways and impairs maturation in plasmacytoiddendritic cells: implications for viral immune escape. Eur J Immunol. 2012 [Epub ahead of print].
11. Benditt EP, Barrett T, McDougall JK. Viruses in the etiology of atherosclerosis. Proc Natl Acad Sci U S A. 1983;80:6386–9.
12. Yamashiroya HM, Ghosh L, Yang R, Robertson Jr AL. Herpesviridae in the coronary arteries and aorta of young trauma victims. Am J Pathol. 1988;130:71–9.
13. Musiani M, Zerbini ML, Muscari A, Puddu GM, Gentilomi G, Gibellini D, Gallinella G, Puddu P, La Placa M. Antibody patterns against cytomegalovirus and Epstein-Barr virus in human atherosclerosis. Microbiologica. 1990;13:35–41.
14. Straka F, Málek I, Staněk V, Ouhrabková R, Gebauerová M, Urbanová D, Vrubel J, Pirk J, Zelízko M, Lánská V. Coronary disease in patients after heart transplantation. Cor Vasa. 1993;35:267–75.
15. Delneste Y, Lassalle P, Jeannin P, Mannessier L, Dessaint JP, Joseph M, Tonnel AB. Production of anti-endothelial cell antibodies by coculture of EBV-infected human B cells with endothelial cells. Cell Immunol. 1993;150:15–26.
16. Tanaka S, Komori K, Okadome K, Sugimachi K, Mori R. Detection of active cytomegalovirus infection in inflammatory aortic aneurysms with RNA polymerase chain reaction. J Vasc Surg. 1994;20:235–43.
17. Luo CY, Ko WC, Tsao CJ, Yang YJ, Su IJ. Epstein-Barr virus-containing T-cell lymphoma and atherosclerotic abdominal aortic aneurysm in a young adult. Hum Pathol. 1999;30:1114–7.
18. Horváth R, Cerný J, Benedík Jr J, Hökl J, Jelínková I, Benedík J. The possible role of human cytomegalovirus (HCMV) in the origin of atherosclerosis. J Clin Virol. 2000;16:17–24.
19. Rupprecht HJ, Blankenberg S, Bickel C, Rippin G, Hafner G, Prellwitz W, Schlumberger W, Meyer J, AutoGene Investigators. Impact of viral and bacterial infectious burden on long-term prognosis in patients with coronary artery disease. Circulation. 2001;104:25–31.

20. Espinola-Klein C, Rupprecht HJ, Blankenberg S, Bickel C, Kopp H, Victor A, Hafner G, Prellwitz W, Schlumberger W, Meyer J. Impact of infectious burden on progression of carotid atherosclerosis. Stroke. 2002;33:2581–6.
21. Espinola-Klein C, Rupprecht HJ, Blankenberg S, Bickel C, Kopp H, Rippin G, Victor A, Hafner G, Schlumberger W, Meyer J, AtheroGene Investigators. Impact of infectious burden on extent and long-term prognosis of atherosclerosis. Circulation. 2002;105:15–21.
22. Shi Y, Tokunaga O. Herpesvirus (HSV-1, EBV and CMV) infections in atherosclerotic compared with non-atherosclerotic aortic tissue. Pathol Int. 2002;52:31–9.
23. De Backer J, Mak R, De Bacquer D, Van Renterghem L, Verbraekel E, Kornitzer M, De Backer G. Parameters of inflammation and infection in a community based case-control study of coronary heart disease. Atherosclerosis. 2002;160:457–63.
24. Khairy P, Rinfret S, Tardif JC, Marchand R, Shapiro S, Brophy J, Dupuis J. Absence of association between infectious agents and endothelial function in healthy young men. Circulation. 2003;107:1966–71.
25. Pitiriga VC, Kotsis VT, Alexandrou ME, Petrocheilou-Paschou VD, Kokolakis N, Zakopoulou RN, Zakopoulos NA. Increased prevalence of Chlamydophila pneumoniae but not Epstein-Barr antibodies in essential hypertensives. J Hum Hypertens. 2003;17:21–7.
26. Pitiriga VC, Kotsis VT, Gennimata V, Alexandrou ME, Papamichail CM, Mitsibounas DN, Petrocheilou-Paschou VD, Zakopoulos NA. Chlamydia pneumoniae and Epstein-Barr antibodies are not associated with carotid thickness: the effect of hypertension. Am J Hypertens. 2003;16:777–80.
27. Andrié R, Braun P, Heinrich KW, Lüderitz B, Bauriedel G. Prevalence of intimal pathogen burden in acute coronary syndromes. Z Kardiol. 2003;92:641–9.
28. Kwon TW, Kim DK, Ye JS, Lee WJ, Moon MS, Joo CH, Lee H, Kim YK. Detection of enterovirus, cytomegalovirus, and Chlamydia pneumoniae in atheromas. J Microbiol. 2004;42:299–304.
29. Ibrahim AI, Obeid MT, Jouma MJ, Moasis GA, Al-Richane WL, Kindermann I, Boehm M, Roemer K, Mueller-Lantzsch N, Gärtner BC. Detection of herpes simplex virus, cytomegalovirus and Epstein-Barr virus DNA in atherosclerotic plaques and in unaffected bypass grafts. J Clin Virol. 2005;32:29–32.
30. Lobzin IV, Boĭtsov SA, Filippov AE, Linchak RM, Mangutov DA. Effect of respiratory infections on the clinical course of coronary artery disease. Klin Med (Mosk). 2005;83:22–6.
31. Schlitt A, Blankenberg S, Weise K, Gärtner BC, Mehrer T, Peetz D, Meyer J, Darius H, Rupprecht HJ. Herpesvirus DNA (Epstein-Barr virus, herpes simplex virus, cytomegalovirus) in circulating monocytes of patients with coronary artery disease. Acta Cardiol. 2005;60:605–10.
32. Lenzi C, Palazzuoli A, Giordano N, Alegente G, Gonnelli C, Campagna MS, Santucci A, Sozzi M, Papakostas P, Rollo F, Nuti R, Figura N. H. pylori infection and systemic antibodies to CagA and heat shock protein 60 in patients withcoronary heart disease. World J Gastroenterol. 2006;12:7815–20.
33. Altun B, Rota S, Demircin M, Reşatoğlu A, Yener A, Bozdayi G. Investigation of herpes group and hepatitis A virus nucleic acids in the atherome plaque samples of patients with coronary arterial disease. Mikrobiyol Bul. 2007;41:537–44.
34. Bayram A, Erdoğan MB, Ekşi F, Yamak B. Demonstration of Chlamydophila pneumoniae, Mycoplasma pneumoniae, Cytomegalovirus, and Epstein-Barr virus in atherosclerotic coronary arteries, nonrheumatic calcific aortic and rheumaticstenotic mitral valves by polymerase chain reaction. Anadolu Kardiyol Derg. 2011;11:237–43.
35. Tremolada S, Delbue S, Ferraresso M, Carloni C, Elia F, Larocca S, Bortolani E, Ferrante P. Search for genomic sequences of microbial agents in atherosclerotic plaques. Int J Immunopathol Pharmacol. 2011;24:243–6.
36. Valmary S, Dorfmüller P, Montani D, Humbert M, Brousset P, Degano B. Human γ-herpes viruses Epstein-Barr virus and human herpesvirus-8 are not detected in the lungs of patients with severe pulmonary arterial hypertension. Chest. 2011;139:1310–6.

37. de Boer OJ, Teeling P, Idu MM, Becker AE, van der Wal AC. Epstein Barr virus specific T-cells generated from unstable human atherosclerotic lesions: implications for plaque inflammation. Atherosclerosis. 2006;184:322–9.
38. Herrero JA, Mathew P, Paya CV. LMP-1 activates NF-kappa B by targeting the inhibitory molecule I kappa B alpha. J Virol. 1995;69:2168–74.
39. McColl SR, Roberge CJ, Larochelle B, Gosselin J. EBV induces the production and release of IL-8 and macrophage inflammatory protein-1 alpha in human neutrophils. J Immunol. 1997;159:6164–8.
40. Waldman WJ, Williams Jr MV, Lemeshow S, Binkley P, Guttridge D, Kiecolt-Glaser JK, Knight DA, Ladner KJ, Glaser R. Epstein-Barr virus-encoded dUTPase enhances proinflammatory cytokine production by macrophages in contact with endothelial cells: evidence for depression-induced atherosclerotic risk. Brain Behav Immun. 2008;22:215–23.
41. Papanicolaou DA, Wilder RL, Manolagas SC, Chrousos GP. The pathophysiologic roles of interleukin-6 in human disease. Ann Intern Med. 1998;128:127–37.
42. Kempe S, Heinz P, Kokai E, Devergne O, Marx N, Wirth T. Epstein-Barr virus-induced gene-3 is expressed in human atheroma plaques. Am J Pathol. 2009;175:440–7.
43. Apostolou F, Gazi IF, Lagos K, Tellis CC, Tselepis AD, Liberopoulos EN, Elisaf M. Acute infection with Epstein-Barr virus is associated with atherogenic lipid changes. Atherosclerosis. 2010;212:607–13.
44. Andrié RP, Bauriedel G, Tuleta I, Braun P, Nickenig G, Skowasch D. Impact of intimal pathogen burden in acute coronary syndromes–correlation with inflammation, thrombosis, and autoimmunity. Cardiovasc Pathol. 2010;19:e205–10.
45. Al-Ghamdi A. Role of herpes simplex virus-1, cytomegalovirus and Epstein-Barr virus in atherosclerosis. Pak J Pharm Sci. 2012;25:89–97.

Chapter 4
The Role of Enteroviruses, Parvovirus B19, Respiratory Syncytial Virus, and Measles Virus in Atherosclerosis and Related Diseases

Abstract The current chapter is aimed to accumulate and discuss existing data on possible association of enteroviruses, parvoviruses, respiratory syncytial virus, and measles virus with atherosclerosis and related diseases. It seems to be that the association of enterovirus infection with atherosclerosis and related diseases is population-dependent. In addition, the evidence that enterovirus genome was detected in the arterial atherosclerotic plaques and endomyocardial tissues by three different research groups in Korea and France may also testify about the association between this virus and atherosclerosis. With respect to other viruses, the results are very scarce, and a number of further investigations is definitely required for the detailed analysis of its role in the etiopathogenesis of cardiovascular diseases. However, parvovirus B19 was detected in arterial specimens of the patients suffered from myocardial infarction. In relation to measles virus, even the opposite (anti-atherosclerotic) effect is admissible.

4.1 Enteroviruses

Enterovirus infections are a significant cause of morbidity and mortality all over the world. The original classification of enteroviruses divided them into the four groups: polioviruses, Coxsackie A viruses, Coxsackie B viruses, and ECHO (Enteric Cytopathic Human Orphan) viruses. A large number of circulating strains in human populations suggested a potential role for these viruses in the development of a number of diseases [1].

The importance of enteroviruses, mainly those of the Coxsackie group, as possible causes of various cardiovascular diseases has been studied from the mid-1970s [2–4]. Heart or vascular damage caused by viral infection might elevate risk of diseases associated with vascular occlusion. In the investigation by Woods et al. [5] 20 (8.6 %) out of 233 Australian patients with transmural myocardial infarction had antibodies against Coxsackie virus B. In the similar study by Nicholls et al. [6] this percentage was

A. Kutikhin et al., *Viruses and Atherosclerosis*, SpringerBriefs in Immunology 4, DOI 10.1007/978-1-4614-8863-7_4, © Springer Science+Business Media New York 2013

threefold higher (10 out of 38 British individuals with acute myocardial infarction, 26.3 %). However, there was no control group in these studies, so their results should be interpreted with caution. In the following study by Wood et al. [7] the control group consisted of subjects with chest pain but without myocardial infarction who were admitted to the same clinic as cases. Coxsackie B virus infection was detected serologically in 7 (13.5 %) out of 52 Scottish cases and in 10 (19 %) out of 52 controls. Authors have also noted that extended investigations over the whole calendar year are unnecessary since their findings agreed with previous failure to reveal a higher frequency of enterovirus infections over a 6-year period among patients with acute myocardial infarction in comparison with those who had non-cardiac diseases [4]. An investigation by El-Hagrassy et al. [8] revealed 4 (13.3 %) out of 30 patients with acute ischaemic heart disease bearing Coxsackie-B-virus-specific IgM.

In 1980, Griffiths et al. [9] demonstrated the identical seropositivity rates and the similar prevalence of raised anti-Coxsackie antibody titres on the sample of 93 patients with myocardial infarction and 99 age- and sex-matched controls from the same geographical area. Two years ago, a study by Lau [10] showed 9.8 % percentage of Coxsackie virus B prevalence among 153 New Zealand individuals with myocardial infarction. The standardized morbidity ratio of Coxsackie B infection was 96 for these subjects in comparison with 104.5 for the control group of patients with diseases other than cardiac ($P > 0.1$). Nevertheless, the standardized morbidity ratio of 91.3 for patients with myocardial infarction was significantly higher than in the second control group (0.0) which consisted of healthy blood donors. In the study carried out by Nikoskelainen et al. [11] 9 (15 %) of 59 Finnish subjects with acute myocardial infarction and 1 (2.6 %) of 38 control patients showed a fourfold or higher antibody increase in paired serum samples against Coxsackie B1–5 viruses, and the difference was statistically significant.

O'Neill et al. [12] assessed the serum samples from 250 subjects suffering from coronary heart disease and 100 control individuals to reveal the presence of anti-Coxsackie B antibodies. The incidence of infection among 130 Scottish patients with acute myocardial infarction was 5 % compared to 4 % in the control group, but in a subgroup of patients with non-transmural myocardial infarction, it reached 14 %. So, an association between Coxsackie infection and myocardial infarction was not found in this investigation. Hannington et al. [13] obtained serum specimens from 105 British patients with myocardial infarction and from 99 age- and sex-matched controls (the patients were the same as in the article of Griffiths et al. [9]). According to their data, ELISA-detected anti-Coxsackie IgM antibodies were revealed in 13 (12.3 %) of the 105 patients with myocardial infarction and in 15 (15.1 %) of the 99 matched controls; therefore, this difference was not statistically significant, and these results confirmed the previous ones obtained by the standard neutralization test. On the contrary, Lau [14] demonstrated that 30 of 153 (19.6 %) New Zealand subjects with acute myocardial infarction and 11 of 178 (6.2 %) matched control blood donors were positive for Coxsackie B virus-specific IgM, and this difference, once again, was statistically significant in New Zealand.

In 1990, Ilbäck et al. [15] conducted a first fundamental study devoted to the role of Coxsackie virus in atherosclerosis. The authors revealed that accumulation of

14C-cholesterol by Balb/c mice increased by 75 % ($P<0.001$) in the heart and by 92 % ($P<0.001$) in the aorta 7 days after CB3 infection. The virus has also caused extensive inflammatory lesions (4.5 % of tissue section area) and lipid accumulation in the myocardium 7 days after inoculation, therefore playing a role in both arterial and myocardial lipid accumulation and acting as initiating factors of atherosclerosis. Seven years later, Conaldi et al. [16] investigated the interactions of Coxsackie B virus with human vascular endothelial cells *in vitro*, founding that the persistence of CVB-3 and -5 correlated with the chronic release of TNF-α, a cytotoxic cytokine which also has a negative inotropic effect on myocardial cells. The authors proposed that both direct viral pathogenicity and indirect cytokine-mediated effects potentially may contribute to vascular damage in the course of systemic Coxsackie B virus infections.

Next year, Roivainen et al. [17] evaluated the presence of IgG antibodies to an enterovirus-common (EVC) antigen and to heat-denatured coxsackievirus B5 in 183 Finnish men and 81 women with myocardial infarction and 379 matched controls. In univariate analysis, EVC antibodies were significantly associated with the risk of myocardial infarction in men ($P=0.009$). Notably, the patients with myocardial infarction had a significantly higher mean level of EVC antibodies than matched controls ($P=0.014$). High antibody levels to EVC were associated with an elevated risk of myocardial infarction in men aged 25–49 years (relative risk$=4.34, P<0.001$) but not in older men (>50 years of age). Next year, Roivainen [18] assessed the association of enterovirus-specific antibodies and cardiovascular events in three large, prospective population nested case-control studies. The study sample consisted of 276 Finnish men with myocardial infarction and their matched control patients. High enterovirus-antibody level was established as an independent risk factor for future cardiovascular events in men aged 25–49 years and in subjects with low level of serum cholesterol. Then, Roivainen et al. [19] measured baseline levels of C-reactive protein (CRP) and antibodies to enterovirus in 241 Finnish individuals with either myocardial infarction or coronary death and in 241 controls without coronary events during the 8.5-year trial in the Helsinki Heart Study. Unexpectedly, the level of antibodies to enterovirus did not differ in cases compared to controls. The next investigation of this research group (Reunanen et al. [20]) included 441 Finnish men with nonfatal myocardial infarction or coronary death within a mean follow-up time of 10 years (276 men without and 165 with major coronary heart disease at baseline), and 840 age-, heart disease status-, and residence-matched controls (305 and 535 controls with and without heart disease, respectively). Subjects without reported baseline heart disease, but not those with heart disease, showing the highest quartile of anti-enterovirus antibodies had a significantly higher risk of coronary events in comparison with men with lower level of antibodies [adjusted OR = 1.71 (95 % CI = 1.09–2.68)]. However, an association was not detected in men with increased serum cholesterol concentration. Possibly, this lack of association may explain the absence of an association in the previous investigation (Roivanen et al. [19]) where only hyperlipidemic subjects were included in the study sample. The authors have also noticed that all the previous studies tested the associations of myocardial infarction with specific enterovirus infections but not with common enterovirus antigen.

Two years later, Kwon et al. [21] investigated the presence of infectious agents in human atherosclerotic arterial plaques. They extracted atherosclerotic plaques from 128 Korean patients with occlusive disease and from 20 normal arterial wall samples, obtained from transplant donors with no history of diabetes, hypertension, smoking, or hyperlipidemia. Enteroviral RNA was found by PCR in 22 (17.2 %) of 128 atherosclerotic vascular lesions but was not detected in any of the control specimens, suggesting a connection of enteroviral infection with atherosclerosis. In the same year, Choy et al. [22] found that post-infection constriction of septal arteries to pressures equal to or less than 60 mm Hg was enhanced in Coxsackie virus B-infected mice compared with sham controls, and at day 42 there was a significant decrease in acetylcholine-induced vasodilation in Coxsackie virus B3-infected mice. Based on their results, the authors proposed that Coxsackie virus infection may be related to cardiovascular disease and essential hypertension. Kis et al. [23] collected blood samples from 59 Hungarian subjects with ischemic stroke and 52 control patients and investigated them on the presence of enterovirus by PCR. Viral RNA was observed neither in cases nor in controls, and the authors revealed no association of stroke with this infectious agent. In the study by Liu et al. [24] on the sample of 488 hypertensive and 942 normotensive subjects Chinese Mongolians, the presence of IgG antibodies against Coxsackie virus was significantly associated with an essential hypertension [76.2 % among hypertensives, 50.3 % among normotensives, OR= 3.17 (95 % CI = 2.31–4.35), $P < 0.001$ after adjustment for risk factors]. Andreoletti et al. [25] detected enterovirus infection markers in 20 (40 %) of 50 French patients who died of myocardial infarction but only in 2 (4 %) of 50 matched subjects without cardiac disease ($P < 0.001$) and only in 4 (8 %) of 50 matched patients with noncoronary chronic cardiopathy ($P < 0.001$). All of the EV RNA-positive patients were positive for capsid viral protein 1 (VP1) indicating viral protein synthesis activity. The *VP1* gene sequences demonstrated a strong homology with sequences of coxsackievirus B2 and B3 serotypes. Immunohistochemical analyses revealed that there was disruption of the sarcolemmal localization of dystrophin in the coxsackievirus-infected tissue areas.

In the investigation by Schanen et al. [26] 4 (22.2 %) of 18 French patients who underwent artery resection were positive for enterovirus genome, suggesting that small amounts of enterovirus genome can be found in lesions of patients with advanced arteriosclerosis. Pesonen et al. [27] revealed that anti-enterovirus IgA titers were significantly higher in 110 Sweden patients suffering from myocardial infarction in comparison with 323 matched controls (OR = 2.72, $P < 0.001$) but do not correlate with the degree of coronary obstruction (Pesonen et al. [28]). Plotkin et al. [29] showed that the relative level of enterovirus antigen (RLEVA) in the blood of patients with myocardial infarction complicated and uncomplicated by cardiogenic shock and/or cardiac rupture was significantly higher than in patients with unstable angina. In addition, concentration of RLEVA in necrotized myocardial areas after death from cardiogenic shock and/or cardiac rupture was higher than in outside myocardial infarction zones, and RLEVA in coronary vessels feeding the necrotic zones of patients with myocardial infarction complicated by cardiogenic shock was higher than in the vessels feeding tissues outside the myocardial infarction zone. The authors suggested that enterovirus is directly involved in the pathogenesis of myocardial infarction and promotes the development of cardiogenic shock and/or cardiac rupture.

4.2 Parvovirus B19

Although the role of human parvovirus B19 in a number of diseases such as erythema infectiosum (fifth disease), aplastic anemia, and certain autoimmune disorders is widely established, its role in atherosclerosis and related diseases was still unexplored until 2009, when Liu et al. [30] recruited 90 Taiwanese patients with coronary artery disease and 475 controls for the assessment of the possible parvovirus B19 role in the atherosclerosis development. They evaluated the titer of the specific IgM and IgG antibodies in all the investigated subjects, revealing that anti-B19 IgG antibodies were observed 1.5- to 2.7-fold more frequently in the individuals with coronary artery disease compared to healthy controls. Anti-B19 IgM antibodies were not detected neither in patients with coronary artery disease nor in healthy controls. According to the nonradioactive in situ PCR, the majority of B19-specific DNA was located in the endothelial cells of the thickened intima. The authors concluded that B19 infection may play a role in the etiopathogenesis of atherosclerosis. Grub et al. [31] compared the frequency of parvovirus B19 infection in 67 Norwegian subjects with both coronary artery disease and inflammatory rheumatic diseases, 52 patients with coronary artery disease alone, and in 30 healthy controls, but detected neither statistically significant difference in the distribution of anti-B19 antibodies nor association between the examined antibodies and the frequency of aortic adventitial mononuclear cell infiltrates. Motor et al. [32] investigated the association of Coxsackie B virus with essential hypertension. In their study, anti-Coxsackie IgM antibodies were found by ELISA in 27 (30 %) out of 90 Turkish subjects with essential hypertension and in 7 (15.6 %) out of 45 controls, whereas IgG antibodies were detected in 27 (30 %) patients and in 14 (31.1 %) control subjects; these differences were not statistically significant. Cases and controls also had the similar levels of serum endotheline-1 and nitric oxide (NO), which represent the antagonistic pair in terms of blood pressure control. In the study by Padmavati et al. [33] parvovirus B was detected by PCR in 8 (8.3 %) out of 98 of tissue specimens. In addition, 3 (6.8 %) of 44 Indian patients with myocardial infarction were also positive for parvovirus B-19 IgG antibodies. However, there was no assessment for the presence of parvovirus B-19 in the control group.

4.3 Respiratory Syncytial Virus (RSV)

Since its first discovery in the 1950s, human respiratory syncytial virus (HRSV) has been recognized as the leading viral pathogen of severe respiratory tract diseases in infants and young children. Certain research groups have also investigated its possible role in the development of atherosclerosis and related diseases. Chang et al. [34] demonstrated that vascular endothelial cells exposed to culture supernatants from RSV-infected pulmonary epithelial A549 cells began to express cell surface molecules, such as ICAM-1, VCAM-1, and E-selectin. Moreover, the authors identified IL-1α as the predominant endothelial cell-activating factor by the pretreatment of epithelial cell supernatants with anti-IL-1α antibody. Guan et al. [35] evaluated the association of previous RSV infection and acute myocardial infarction. In their study, patients with myocardial infarction had anti-RSV IgG antibodies significantly more frequently than controls (adjusted OR = 11.1, 95 % CI = 3.3–29.5). So, the study supported the hypothesis that the previous RSV infection can be associated with myocardial infarction.

4.4 Measles Virus

A potential impact of measles virus on the development of atherosclerosis was first investigated by Csonka et al. [36] The authors investigated *in vitro* effect of the measles virus on the aortic endothelial and smooth muscle cells. The virus-infected aortic cells occasionally formed syncytii and contained nuclear inclusions, whereas virus-infected endothelial cells lysosome contained viral nucleocapsids. The early phase of measles virus replication inhibited the proliferation of endothelial cells, simultaneously stimulating the replication of the smooth muscle cells. The authors concluded that measles virus infection can be a cause of atherosclerosis, damaging endothelial cells by altering the cell membrane permeability and inducing proliferation of aortic smooth muscle cells. As opposed to this investigation, Ait-Oufella et al. [37] hypothesized that the anti-inflammatory properties of measles virus nucleoprotein may hinder the development of atherosclerosis. In their research, permanent administration of measles virus nucleoprotein to apolipoprotein E-deficient mice initiated an anti-inflammatory T-regulatory-cell type 1-like response and impeded macrophage and T-cell accumulation within the lesions. Moreover, treatment by measles virus nucleoprotein significantly reduced the development of new atherosclerotic plaques and inhibited the progression of established lesions. The anti-atherosclerotic potential of nucleoprotein was retained in its short N-terminal segment, and the protective effects were lost in mice with lymphocyte deficiency.

4.5 Discussion

In this chapter, the whole body of the available data on the association of enterovirus (mainly Coxsackie B virus) with atherosclerosis and related diseases was analyzed. All the epidemiological articles discussing the potential association of viruses considered in this review with atherosclerosis and related diseases are summarized in Table 4.1. First of all, it seems to be that the association is somehow population-dependent.

Table 4.1 The association of various viruses with atherosclerosis, coronary artery disease, acute coronary syndrome, angina, myocardial infarction, arterial hypertension, and stroke

Authors, reference, population	Method of virus detection, sample source	Number of cases and controls	Virus prevalence among cases and controls	Association of virus with atherosclerosis
Enterovirus				
Woods et al. [5] Australian population	Serology Serum	233 cases with myocardial infarction	8.6 %	No control group
Nicholls et al. [6] British population	Serology Serum	38 cases with myocardial infarction	26.3 %	No control group
Wood et al. [7] Scottish population	Assay for neutralizing antibodies to the six types of Coxsackie B virus by a micro-metabolic inhibition test (microneutralization test) Serum	52 cases with myocardial infarction, 52 controls with chest pain but without myocardial infarction	13.5 % in cases, 19 % in controls	No association
El-Hagrassy et al. [8]	ELISA IgM antibodies Serum	30 cases with acute ischaemic heart disease	13.3 %	No control group
Griffiths et al. [9] British population	Serology Serum	93 cases with myocardial infarction, 99 controls	3.2 % in cases, 2 % in controls	No association
Lau [10] New Zealand population	Serology Serum	153 cases with myocardial infarction	9.8 %	SMR in cases = 96 SMR in control group of patients with diseases other than cardiac = 104.5 $P>0.1$ SMR in patients with myocardial infarction = 91.3 SMR in control group of healthy blood donors = 0.0 Positive association

(continued)

Table 4.1 (continued)

Authors, reference, population	Method of virus detection, sample source	Number of cases and controls	Virus prevalence among cases and controls	Association of virus with atherosclerosis
Nikoskelainen et al. [11] Finnish population	Microneutralization test Serum	59 cases with myocardial infarction, 38 controls	15 % in cases, 2.6 % in controls	Positive association
O'Neill et al. [12] Scottish population	Microneutralization test Serum	130 cases with myocardial infarction, 100 controls	5 % in cases, 4 % in controls	No association
Hannington et al. [13] British population	ELISA IgM antibodies Serum	105 cases with myocardial infarction, 99 controls (the same as in the study of Griffiths et al.)	12.3 % in cases, 15.1 % in controls	No association
Lau [14] New Zealand population	ELISA IgM antibodies Serum	153 cases with myocardial infarction, 178 controls	19.6 % in cases, 6.2 % in controls	Positive association
Roivainen et al. [17] Finnish population	ELISA IgG antibodies Serum	264 cases with myocardial infarction, 379 controls		Positive association
Roivainen [18] Finnish population	ELISA IgG antibodies Serum	276 cases with myocardial infarction, 276 controls		Positive association among men aged 25–49 years and in men with low level of serum cholesterol
Roivainen et al. [19] Finnish population	ELISA IgG antibodies	241 cases with myocardial infarction or coronary death, 241 controls		No association
Reunanen et al. [20] Finnish population	ELISA IgG antibodies	441 cases with myocardial infarction or coronary death, 840 controls, mean follow-up of 10 years		Positive association Subjects with heart disease at baseline: OR = 1.71 (95 % CI = 1.09–2.68)
Kwon et al. [21] Korean population	PCR Arterial atherosclerotic plaques	128 cases with occlusive disease, 20 controls	17.2 % in cases, 0 % in controls	Positive association

Kis et al. [23] Hungarian population	PCR Serum	59 cases with ischemic stroke, 52 controls	0 % in both cases and controls	No association
Liu et al. [24] Chinese Mongolian population	ELISA IgG antibodies Serum	488 hypertensive cases and 942 normotensive controls	76.2 % in cases, 50.3 % in controls	Positive association OR = 3.17 (95 % CI = 2.31–4.35)
Andreoletti et al. [25] French population	PCR Endomyocardial tissues	50 cases with myocardial infarction, 50 controls without cardiac disease, 50 controls with noncoronary chronic cardiopathy	40 % in cases, 4 % in controls without cardiac disease, 8 % in controls with noncoronary chronic cardiopathy	Positive association
Schanen et al. [26] French population	PCR Arterial atherosclerotic plaques	18 cases after the artery resection	22.2 %	No control group
Pesonen et al. [27] Sweden population	ELISA IgA antibodies Serum	110 cases with myocardial infarction, 323 controls		Positive association OR = 2.72
Parvovirus Liu et al. [30] Taiwanese population	ELISA IgM and IgG antibodies Serum	90 cases with coronary artery disease, 475 controls		Positive association
Grub et al. [31] Norwegian population	ELISA Serum	67 cases with both coronary artery disease and inflammatory rheumatic diseases, 52 cases with coronary artery disease alone, 30 controls		No association

(continued)

Table 4.1 (continued)

Authors, reference, population	Method of virus detection, sample source	Number of cases and controls	Virus prevalence among cases and controls	Association of virus with atherosclerosis
Motor et al. [32] Turkish population	ELISA IgM and IgG antibodies Serum	90 cases with essential hypertension, 45 controls	IgM antibodies: 30 % in cases, 15.6 % in controls IgG antibodies: 30 % in cases, 31.1 % in controls	No association
Padmavati et al. [33] Indian population	PCR, ELISA IgG antibodies Arterial specimens, serum	98 cases with myocardial infarction	8.3 % in tissue specimens of cases, 6.8 % in blood of cases	No assessment in the control group
Respiratory syncytial virus				
Guan et al. [35] Chinese population	ELISA IgG antibodies Serum			Positive association

ELISA enzyme-linked immunosorbent assay, *Ig* immunoglobulin, *SMR* standard mortality ratio, *OR* odds ratio, *CI* confidence interval, *PCR* polymerase chain reaction

Two studies in New Zealand [10, 14] and four investigations in Finland [17–20] proved that there is a significantly positive association in these countries, and the similar results were obtained in Sweden [27] as well. Two investigations carried out in China [24] and Korea [21] also found the same positive association, pointing out the dependence of the positivity of the association on geographical location. The analogical studies in Denmark and Norway (and/or in other Asian countries), possibly, might shed light on this issue. In contrast, all four studies in Great Britain [6, 7, 12, 13] (two of them in Scotland) [7, 12] did not reveal any association between enterovirus and myocardial infarction, feasibly due to the population features in this geographical region. It is also possible that the sample size was insufficient in the studies where the negative result was obtained (the largest one did not exceed 130 cases and 100 controls), or enterovirus does not play a role in stroke development. In addition, the evidence that enterovirus genome was detected in the arterial atherosclerotic plaques and endomyocardial tissues by three different research groups in Korea [21] and France [25, 26] may also testify about the association between this virus and atherosclerosis.

It is important to note that the determination of the whole spectrum of anti-Coxsackie antibodies (IgA, IgM and IgG) can be recommended for the forthcoming studies, since all of them were used in previous ones. The method of PCR is rather widespread, but it is usually used for the detection of the viral genome in arterial atherosclerotic plaques or endomyocardial tissues but not in the serum. The enlargement of the sample size and a conduction of multicenter studies are worthwhile as well. In the relation to the distinct pathologies, only the association with myocardial infarction was evaluated in detail. Concerning other ones, there was one investigation revealing a positive association between enterovirus infection and arterial hypertension, and one study failed to find a connection of enterovirus with stroke. So, the assessment of enterovirus influence on the frequency of other occlusive pathologies is the issue for further studies.

With respect to other viruses, the results are very scarce. Although in one Taiwanese [30] investigation the positive association of parvovirus B19 with coronary artery disease was detected, the Norwegian [31] study did not confirm these results. In addition, the Turkish [32] study devoted to the role of parvovirus B19 in the essential hypertension development did not find an association between them; however, these Norwegian and Turkish studies had rather small sample size (not more than 90 cases and 45 controls), and their results should be interpreted with caution. It is also worth a note that parvovirus B19 was detected in arterial specimens of the patients suffered from myocardial infarction. However, at the moment a lack of relevant investigations does not allow to establish this virus as a causative agent of atherosclerosis or vascular occlusion-related diseases.

Although the results of the only epidemiological study indicating the impact of respiratory syncytial virus on the risk of atherosclerosis are rather promising and the odds ratio is certainly high (more than 11), a number of further investigations are definitely required for the detailed analysis of its role in the etiopathogenesis of cardiovascular diseases.

At the moment, it is feasible to suggest that enterovirus may be one of the possible causes of atherosclerosis and related diseases such as myocardial infarction, although this hypothesis has certain shortcomings and contradictions, and additional basic and epidemiological studies are required to prove it. Regarding the parvovirus B19 and respiratory syncytial virus, the data are insufficient for making any clear conclusions. In relation to measles virus, even the opposite (anti-atherosclerotic) effect is admissible.

References

1. Victoria JG, Kapoor A, Li L, Blinkova O, Slikas B, Wang C, et al. Metagenomic analyses of viruses in stool samples from children with acute flaccid paralysis. J Virol. 2009;83(9): 4642–51.
2. Bell EJ, Grist NR. Coxsackievirus infections in patients with acute cardiac disease and chest pain. Scott Med J. 1968;13:47–51.
3. Dawson KP, Rogen AS. Cardiac complications of Coxsackie virus group B infection. Practitioner. 1970;205:333–5.
4. Grist NR, Bell EJ. A six-year study of Coxsackievirus B infections in heart disease. J Hyg. 1974;73:165–72.
5. Woods JD, Nimmo MJ, Mackay-Scollay EM. Acute transmural myocardial infarction associated with active Coxsackie virus B infection. Am Heart J. 1975;89(3):283–7.
6. Nicholls AC, Thomas M. Coxsackie virus infection in acute myocardial infarction. Lancet. 1977;1(8017):883–4.
7. Wood SF, Rogen AS, Bell EJ, Grist NR. Role of Coxsackie B viruses in myocardial infarction. Br Heart J. 1978;40(5):523–5.
8. El-Hagrassy MM, Banatvala JE, Coltart DJ. Coxsackie-B-virus-specific IgM responses in patients with cardiac and other diseases. Lancet. 1980;2(8205):1160–2.
9. Griffiths PD, Hannington G, Booth JC. Coxsackie B virus infections and myocardial infarction. Results from a prospective, epidemiologically controlled study. Lancet. 1980;1(8183): 1387–9.
10. Lau RC. Coxsackie B, virus infection in acute myocardial infarction and adult heart disease. Med J Aust. 1982;2(11):520–2.
11. Nikoskelainen J, Kalliomäki JL, Lapinleimu K, Stenvik M, Halonen PE. Coxsackie B virus antibodies in myocardial infarction. Acta Med Scand. 1983;214(1):29–32.
12. O'Neill D, McArthur JD, Kennedy JA, Clements G. Coxsackie B virus infection in coronary care unit patients. J Clin Pathol. 1983;36(6):658–61.
13. Hannington G, Booth JC, Bowes RJ, Stern H. Coxsackie B virus-specific IgM antibody and myocardial infarction. J Med Microbiol. 1986;21(4):287–91.
14. Lau RC. Coxsackie B, virus-specific IgM responses in coronary care unit patients. J Med Virol. 1986;18(2):193–8.
15. Ilbäck NG, Mohammed A, Fohlman J, Friman G. Cardiovascular lipid accumulation with Coxsackie B virus infection in mice. Am J Pathol. 1990;136(1):159–67.
16. Conaldi PG, Serra C, Mossa A, Falcone V, Basolo F, Camussi G, et al. Persistent infection of human vascular endothelial cells by group B coxsackieviruses. J Infect Dis. 1997;175(3): 693–6.
17. Roivainen M, Alfthan G, Jousilahti P, Kimpimäki M, Hovi T, Tuomilehto J. Enterovirus infections as a possible risk factor for myocardial infarction. Circulation. 1998;98(23):2534–7.
18. Roivainen M. Enteroviruses and myocardial infarction. Am Heart J. 1999;138(5 Pt 2): S479–83.
19. Roivainen M, Viik-Kajander M, Palosuo T, Toivanen P, Leinonen M, Saikku P, et al. Infections, inflammation, and the risk of coronary heart disease. Circulation. 2000;101(3):252–7.

20. Reunanen A, Roivainen M, Kleemola M, Saikku P, Leinonen M, Hovi T, et al. Enterovirus, mycoplasma and other infections as predictors for myocardial infarction. J Intern Med. 2002;252(5):421–9.

21. Kwon TW, Kim DK, Ye JS, Lee WJ, Moon MS, Joo CH, et al. Detection of enterovirus, cytomegalovirus, and Chlamydia pneumoniae in atheromas. J Microbiol. 2004;42(4):299–304.

22. Choy JC, Lui AH, Moien-Afshari F, Wei K, Yanagawa B, McManus BM, et al. Coxsackievirus B3 infection compromises endothelial-dependent vasodilation of coronaryresistance arteries. J Cardiovasc Pharmacol. 2004;43(1):39–47.

23. Kis Z, Sas K, Gyulai Z, Treso B, Petrovay F, Kapusinszky B, et al. Chronic infections and genetic factors in the development of ischemic stroke. New Microbiol. 2007;30(3):213–20.

24. Liu L, Liu Y, Tong W, Ye H, Zhang X, Cao W, et al. Pathogen burden in essential hypertension. Circ J. 2007;71(11):1761–4.

25. Andréoletti L, Ventéo L, Douche-Aourik F, Canas F, Lorin de la Grandmaison G, Jacques J, et al. Active Coxsackieviral B infection is associated with disruption of dystrophin in endomyocardial tissue of patients who died suddenly of acute myocardial infarction. J Am Coll Cardiol. 2007;50(23):2207–14.

26. Schanen C, Nasri D, Bourlet T, Barral X, Favre JP, Bourrat D, et al. Enterovirus in arteriosclerosis: a pilot study. J Clin Virol. 2007;39(2):106–12.

27. Pesonen E, Andsberg E, Grubb A, Rautelin H, Meri S, Persson K, et al. Elevated infection parameters and infection symptoms predict an acute coronary event. Ther Adv Cardiovasc Dis. 2008;2(6):419–24.

28. Pesonen E, El-Segaier M, Persson K, Puolakkainen M, Sarna S, Ohlin H, et al. Infections as a stimulus for coronary occlusion, obstruction, or acute coronary syndromes. Ther Adv Cardiovasc Dis. 2009;3(6):447–54.

29. Plotkin VI, Voronel' VL, Timoshina MA, Zaripova ZA, Murina EA, Khromov-Borisov NN. Enterovirus infection as a risk factor of acute coronary syndrome and its complications. Klin Med (Mosk). 2011;89(2):25–9.

30. Liu SC, Tsai CT, Wu CK, Yu MF, Wu MZ, Lin LI, et al. Human parvovirus b19 infection in patients with coronary atherosclerosis. Arch Med Res. 2009;40(7):612–7.

31. Grub C, Brunborg C, Hasseltvedt V, Aukrust P, Førre O, Almdahl SM, et al. Antibodies to common infectious agents in coronary artery disease patients with and without rheumatic conditions. Rheumatology (Oxford). 2012;51(4):679–85.

32. Motor VK, Arica S, Motor S, Yilmaz N, Evirgen O, Inci M, et al. Investigation of parvovirus B19 seroprevalence, endothelin-1 synthesis, and nitric oxide levels in the etiology of essential hypertension. Clin Exp Hypertens. 2012;34(3):217–21.

33. Padmavati S, Gupta U, Agarwal HK. Chronic infections and coronary artery disease with special reference to Chalmydia pneumoniae. Indian J Med Res. 2012;135:228–32.

34. Chang CH, Huang Y, Anderson R. Activation of vascular endothelial cells by IL-1alpha released by epithelial cells infected with respiratory syncytial virus. Cell Immunol. 2003; 221(1):37–41.

35. Guan XR, Jiang LX, Ma XH, Wang LF, Quan H, Li HY. Respiratory syncytial virus infection and risk of acute myocardial infarction. Am J Med Sci. 2010;340(5):356–9.

36. Csonka E, Bayer PI, Büki K, Várady G. Influence of the measles virus on the proliferation and protein synthesis of aortic endothelial and smooth muscle cells. Acta Microbiol Hung. 1990; 37(2):193–200.

37. Ait-Oufella H, Horvat B, Kerdiles Y, Herbin O, Gourdy P, Khallou-Laschet J, et al. Measles virus nucleoprotein induces a regulatory immune response and reduces atherosclerosis in mice. Circulation. 2007;116(15):1707–13.

Chapter 5
Hepatitis Viruses, Atherosclerosis, and Related Diseases

Abstract In this chapter, a role of various hepatitis viruses in the development of atherosclerosis will be discussed. At the moment, it is hard to conclude definitely whether HAV is associated with atherosclerosis and related diseases but it seems to be that this correlation is population-dependent, and may be based on the causation of endothelial dysfunction. Regarding HBV, the reliable association of the chronic infection caused by this virus with atherosclerosis and related diseases was not revealed. Finally, it was demonstrated that chronic HCV infection is associated with subclinical atherosclerosis, CAD, and stroke, most probably through the direct inflammatory mechanism.

5.1 Hepatitis A Virus (HAV), Atherosclerosis, and Related Diseases

There are only a few studies investigating the possible association of HAV with atherosclerosis and related diseases. Zhu et al. [1] were the first who studied the possible association between hepatitis A virus (HAV) infection and coronary artery disease (CAD). In their investigation, of the 391 US patients, CAD prevalence was 74 % in HAV-seropositive (IgG antibodies) and 52 % in HAV-seronegative patients. In addition, C-reactive protein (CRP) levels were significantly higher in HAV-seropositive than in HAV-seronegative patients. Logistic regression analysis demonstrated that HAV seropositivity is an independent predictor of risk for CAD and elevated CRP levels. Therefore, authors concluded that HAV infection is associated with CAD, and suggested that this virus may play a causal role in atherogenesis. The same research group reported a 1.6-fold increased risk (95 % CI = 1.1–2.3) for myocardial infarction or death among individuals positive for IgG antibodies against HAV [2]. These results were further confirmed by Prasad et al. [3] who also established HAV as a possible causative agent of CAD (odds ratio [OR] = 2.6, 95 % confidence interval [CI] = 1.4–5). In addition, HAV infection was an independent predictor of endothelial dysfunction, determined as the percent change in coronary vascular resistance in response to acetylcholine but not the responses to sodium nitroprusside or adenosine [3]. Moreover, HAV infection was also an independent determinant of endothelial function in the subgroup with angiographically normal coronary arteries [3].

In contrast, Cainelli et al. [4] considered that the epidemiological evidence argues against a significant role for HAV infection in atherogenesis. According to their opinion, HAV infection is highly prevalent in Saudi Arabia, Yemen [5], Egypt [6], and sub-Saharan Africa, and its prevalence has been reduced only recently in southern Italy [7]. In contrast, HAV infection is far less frequent in northern European countries and in Australia [8]. However, if HAV infection has a significant role in atherogenesis, the pattern of the incidence of atherosclerosis manifestations would be similar to that for HAV infection, but this is not the case. Since the 1970s, a very low incidence of atherosclerotic diseases has been noted in southern Italy [9], and developing regions have a relatively low burden of ischemic heart disease and stroke; in contrast, the incidence of cardiovascular diseases is remarkably higher in Europe, North America, and Australia [10]. Nevertheless, these arguments can be refuted due to the multifactorial nature of atherosclerosis, and HAV may be only one out of a number of its causes. However, in the further study of Smieja et al. [11] on 107 Canadian subjects either with myocardial infarction or unstable angina and 107 controls, no association of HAV infection with cardiovascular events was revealed.

In the next year, Burnett et al. [12] demonstrated the effect of HAV vaccination on atherosclerosis development in a cholesterol-fed mouse model; nevertheless, after 15 weeks, no significant differences were found in lesion area between the vaccinated and non-vaccinated animals. Furthermore, Auer et al. [13] tested the blood from 218 Austrian patients with CAD for serum IgG antibodies to HAV. In this study, CAD prevalence was 66.3 % in HAV-seropositive and 57.5 % in HAV-seronegative patients, so HAV seropositivity did not influence CAD risk. Similar results were obtained by Ongey et al. in Germany [14] who found that HAV seroprevalence is associated neither with the prevalence of cardiovascular diseases nor with mean levels of blood lipids among patients suffering from diabetes mellitus. Finally, Smieja et al. [15] did not find any association of anti-HAV IgG antibodies with risk of cardiovascular events in Heart Outcomes Prevention Evaluation (HOPE) which included over 3,000 Canadian patients.

So, the data about the role of HAV in the development of atherosclerosis and related diseases are rather scarce. Chronologically, the first three studies found an association of HAV with CAD and myocardial infarction; however, four investigations carried out later did not reveal such correlation. It is interesting that the positive result was obtained only among US population but not among Canadian, Austrian, and German populations, possibly reflecting the population dependence for the association of HAV with CAD and cardiovascular events. Feasibly, certain genetic, environmental, or socioeconomical factors may play a role in this difference; features of viral strains in distinct populations are also plausible. There were no study providing a molecular basis for this association; nevertheless, it was shown that HAV infection statistically significantly correlated with endothelial dysfunction since microvascular and epicardial dilation with acetylcholine tended to be lower in HAV-seropositive subjects compared to HAV-seronegative individuals. It seems to be that inflammation caused by persistent HAV infection, autoimmune reaction to host antigens similar with HAV antigens (so-called phenomenon of molecular mimicry), or persistent infection caused by any pathogen bearing antigens similar to HAV antigens might be three probable mechanisms providing the changes leading to the endothelial dysfunction. However, the exact mechanisms are unclear, and further basic investigations in this field are needed.

5.2 Hepatitis B Virus (HBV), Atherosclerosis, and Related Diseases

A possible role of chronic hepatitis B infection in the pathogenesis of carotid arteriosclerosis was first revealed in cross-sectional cohort study of Ishizaka et al. [16]. Of the 4,686 Japanese subjects, 1,294 (28 %) had carotid artery plaques, 40 (0.9 %) were positive for HBsAg, and HBsAg positivity was positively associated with carotid atherosclerosis with an OR of 1.57 (95 % CI = 1.1–2.24). In contrast, Bilora et al. [17] showed that chronic viral hepatitis may help to prevent atherosclerosis. They compared 48 Italian patients with a histological diagnosis of chronic viral hepatitis (42 hepatitis C virus-related, 6 hepatitis B virus-related) with a low degree of activity and preserved hepatic function, and 50 matched controls, observing clearly lower prevalence of carotid atherosclerosis among patients. According to their data, percentage of patients with carotid atherosclerosis was 27 % in patients with liver disease and 56 % in controls. Moreover, patients with liver disease had fewer atheromatous lesions (16 plaques versus 59) than controls, and they were characterized by a lower degree of vessel stenosis. In addition, negative result was found in a cross-sectional study of Völzke et al. [18] in Germany who revealed that there was no independent association between anti-HBs and anti-HCV antibody seropositivity and atherosclerotic end-points such as prevalent myocardial infarction, stroke, carotid intima-media thickness (IMT), carotid plaques and stenosis. Su et al. [19] were the first who evaluated the effects of chronic HBV infection on serum lipid profile. They detected that among patients with asymptomatic chronic HBV infection, levels of total cholesterol (TC) and high-density lipoprotein cholesterol (HDL-C) were decreased by 5.8 and 2.7 mg/dL, respectively, indicating the possible HBV role in development of dyslipidemia; nevertheless, no association of HBV infection with atherosclerosis was detected in this study [19]. In the investigation of Tong et al. [20], 77 % (224 of 291) of Chinese patients with CAD and 73.4 % (105 of 143) of controls without angiographic evidence of atherosclerosis were seropositive for HBV, so HBV infection did not correlate with CAD risk; however, CRP level was significantly lower in HBV seropositive population. Similar results were revealed by Ghotaslou et al. [21] in their study on over 5,000 Egyptian subjects; the prevalence of HBsAg positivity tended to be higher in CAD patients than in those without CAD (3.28 % vs 2.17 %), but the difference was not statistically significant. In addition, Yang et al. [22] failed to detect any association of HBV infection with carotid atherosclerosis in Taiwanese population. In the same year, Sung et al. [23] investigated an association between HBsAg seropositivity and cardiovascular diseases in a cohort of Koreans included 521,421 individuals, and in this study HBsAg seropositivity was associated with a decreased risk of ischemic stroke and myocardial infarction (OR = 0.79, 95 % CI = 0.68–0.90 and OR = 0.74, 95 % CI = 0.62–0.87, respectively); however, an increased risk of hemorrhagic stroke was detected (OR = 1.33, 95 % CI = 1.15–1.52).

In 2010, Turhan et al. [24] tested 260 inactive HBsAg Turkish carriers and 80 healthy control subjects for the possible association between HBsAg positivity and mean platelet volume (MPV), a newly emerging risk factor for atherothrombosis. The MPV level was significantly higher in the inactive HBsAg carrier group than in the control group (8.8 ± 1.2 fl vs. 8.1 ± 0.9 fl), suggesting that chronic hepatitis B patients with inactive disease tend to have relatively increased platelet activation, and an atherothrombotic risk [24]. However, this study investigated the association of HBsAg carriage with just a risk factor of atherosclerosis, but not with atherosclerosis itself. Wang et al. [25] followed-up 22,472 Taiwanese subjects, consisting of 18,541 HBsAg seronegatives and 3,931 seropositives, for 17 years, and HBsAg seropositivity was not associated with increased mortality risks of atherosclerosis-related diseases. So, these results confirmed the previous ones from Taiwan where the negative result was obtained [19, 22]. The study of Grab et al. [26] in Norway revealed an absence of association of HBV with CAD in patients with rheumatic diseases, however, it may be explained by the small sample size in this study (67 patients with inflammatory rheumatic diseases, 52 patients without IRD and 30 healthy controls). Finally, D:A:D Cohort Study did not find any association between HBV coinfection and the development of myocardial infarction among HIV-infected individuals. Event rates per 1,000 person-years in those who were HBV-seronegative, had inactive infection or had active infection were 3.2 (95 % CI = 2.8–3.5), 4.2 (95 % CI = 3.1–5.2) and 2.8 (95 % CI = 1.8–3.9), respectively. After adjustment, there was no association between inactive HBV infection (rate ratio 1.07, 95 % CI = 0.79–1.43) or active HBV infection (rate ratio 0.78, 95 % CI = 0.52–1.15] and the development of myocardial infarction [27].

It is worth to note that only 2 studies out of 12 revealed a statistically significant association of HBV infection with altered risk of cardiovascular diseases, and the results contained in them are disparate: one study investigated the association of HBV with carotid atherosclerosis, whereas another one aimed to reveal a correlation with myocardial infarction and stroke. Possibly, a population-dependent effect may exist, since there were no studies showing an absence of association in Japan or Korea, however, the exact explanation of the discrepancies between distinct investigations is unclear. In addition, Sung et al. [23] who carried out one of two studies with positive result suggested that HBV infection does not play an important role in the etiology of cardiovascular events through a proinflammatory effect. According to their conclusions, reduced ability to coagulation due to HBV-associated chronic liver dysfunction elevates the risk of hemorrhagic stroke while decreasing the risk of ischemic stroke and myocardial infarction. Significantly lower CRP level among HBV infected patients found in the study of Tong et al. [20] also contradicts the hypothesis of HBV inflammatory role in the development of cardiovascular pathologies. The decreased level of serum total cholesterol also refutes the role of the changes in atherotic profile in possible HBV-related mechanisms of causing cardiovascular diseases. So, at the moment, it is not possible to establish HBV as a causative agent of atherosclerosis and related diseases, despite certain premises, since 10 of 12 studies found no such association, but it is worth to be noted that chronic HBV infection may possibly increase risk of hemorrhagic stroke simultaneously decreasing the risk of ischemic stroke and myocardial infarction, at least in certain populations.

5.3   Hepatitis C Virus (HCV), Atherosclerosis, and Related Diseases

5.3.1   HCV and Changes in Immunological and Atherotic Profile

Community-based study conducted in Egypt has shown that in a single population, chronic HCV infection is associated with glucose intolerance and, despite that, with a favourable lipid pattern, consisting of an elevation in HDL cholesterol and a reduction in total cholesterol (TC), low-density lipoprotein cholesterol (LDL-C) and triglycerides (TGs). An intriguing finding was the high TG level observed among subjects with past infection, suggesting that elevated TGs at the time of acute infection may facilitate viral clearance [28]. In the study of Boddi et al. [29], positive-strand HCV RNA was detected in seven carotid plaque tissues from HCV-positive patients and was not detected in the nine carotid plaque tissues obtained from HCV-negative patients. In three patients, HCV RNA was found in carotid plaque and not in serum. HCV replicative intermediates were detected in three plaque samples. Direct sequencing of HCV RNA from the plaques and serum showed HCV genotypes 2 (five cases) and 1 (two cases).The finding of HCV RNA sequences in plaque tissue strongly suggests an active local infection, and, in turn, confirms the role of this virus in carotid atherosclerosis [29]. Masia et al. investigated the influence of HCV therapy with pegylated interferon-a plus ribavirin on cardiovascular disease risk [30]. A total of 56 patients were included; 32 (57.1 %) were HCV/HIV coinfected and 24 (42.9 %) were HCV monoinfected [30]. Compared with baseline, during HCV therapy there was a significant decrease in the concentrations of matrix metalloproteinase-9 (MMP-9), intercellular cell adhesion molecule-1 (ICAM-1) and oxidized low-density lipoproteins [30]. In contrast, levels of vascular cell adhesion molecule-1 (VCAM-1), monocyte chemotactic protein-1 and fibrinogen increased during treatment [30]. After treatment discontinuation, levels of ICAM-1, VCAM-1 and TNF-α were significantly lower compared with baseline, a change restricted to patients with sustained virological response [30]. Changes in biomarkers were similar in HIV-infected and -uninfected patients [30]. Authors concluded that treatment for HCV induces different changes in several cardiovascular risk biomarkers, most being anti-atherogenic effects, although only the anti-atherogenic effects remained after treatment discontinuation in patients with sustained virological response [30]. In their second study, this research group found that median soluble vascular CAM-1 (sVCAM-1) and intercellular CAM-1 (sICAM-1) levels were significantly higher in HIV/HCV-coinfected patients; however, carotid IMT did not differ between HCV/HIV-coinfected and HIV-monoinfected patients [31]. The study of Oliveira et al. [32] revealed that HCV patients are characterized by higher levels of proinflammatory cytokines (IL-6 and TNF-α) compared to controls, and the relation of proinflammatory/anti-inflammatory TNF-α/IL10 and IL-6/IL10 were higher in HCV patients. The Framingham score directly correlated to IL-6 and TNF-α, but differences were not statistically significant [32]. According to the opinion of Perrin-Cocon et al. [33], in patients chronically infected, HCV interferes with lipoprotein metabolism resulting in the production of infectious modified lipoproteins, which are modified low-density lipoproteins containing viral material that can alter maturation of dendritic cells and affect specific Toll-like receptor signaling.

5.3.2 HCV and Subclinical Atherosclerosis

The association of HCV infection with atherosclerosis and related diseases was first investigated in 2003 by Ishizaka et al. [34], who found HCV core protein being positively associated with carotid atherosclerosis (OR = 5.61, 95 % CI = 2.06–15.26) in the study including almost 2,000 individuals. Four years later, the similar positive results were obtained by Targher et al. [35] who compared carotid intima-media thickness among 60 patients with non-alcoholic steatohepatitis (NASH), 35 and 60 patients with chronic hepatitis B or C, respectively, and 60 control subjects. Sawayama et al. [36] revealed that both chronic HCV infection can reduce the effectiveness of lipid-lowering therapy for carotid atherosclerosis in terms of change of maximum common carotid artery intima-media thickness (Max-IMT). In their study, patients without HCV infection showed a significant reduction of Max-IMT, whilst only a small decrease of Max-IMT was noted in the patients with HCV infection. An opposite result was noted in the study of Caliskan et al. [37] in hemodialysis patients but the study sample was very small (only 72 patients), and since hemodialysis patients had a large number of uremia-related cardiovascular risk factors, the effect of HCV infection could disappear in this group of patients. The investigation of Tien et al. [38] did not find an association of HCV infection with carotid atherosclerosis, but their study sample included only HIV-infected patients that may distort the results creating a bias. In the study of Miyajima et al. [39] IMT, LDL-C and TG levels were significantly lower in the group with chronic infection than the levels in the other groups. In addition, an association with severe insulin resistance and with mild atherosclerosis, suggesting a unique characteristic of HCV-related metabolic abnormality, was revealed [39].

These results were further confirmed in the study of Sosner et al. [40], in which prevalence of subclinical carotid atherosclerosis was significantly higher in HCV-HIV co-infected patients compared to HIV-monoinfected patients (for HCV infection OR = 10, 95 % CI = 1.5–72), despite of the fact that LDL-C and blood pressure (BP) were lower in this group.

5.3.3 HCV and CAD

Vassalle et al. [41] evaluated whether seropositivity for HCV is associated with the CAD occurrence. They recruited 491 patients and 195 controls, founding HCV seropositivity to be associated with the presence of CAD (OR = 3.2, 95 % CI = 1.1–9.2) [41]. These results were affirmed by Tsui et al. [42], who found HCV-seropositive patients having higher rates of death, cardiovascular events, and heart failure hospitalizations during follow-up. In addition, these subjects had also significantly lower mean levels of CRP and fibrinogen but higher levels of TNF-α [42]. In the large study of Butt et al. [43], which included 82,083 HCV-infected and 89,582 HCV-uninfected subjects, HCV infected subjects were less likely to have hypertension, hyperlipidemia, and had lower total plasma cholesterol, LDL-C, and TG levels compared to HCV-uninfected subjects. However, despite a favorable risk profile, HCV infection was also associated with a 1.25-fold higher risk of CAD (95 % CI = 1.2–1.3) [43]. The study of Grab et al. [26] revealed an absence of association of HCV with CAD in patients with rheumatic diseases, however, it may be explained by the small sample size in this study (67 patients with inflammatory rheumatic diseases, 52 patients without IRD and 30 healthy controls).

5.3.4 HCV and Myocardial Infarction

Arcari et al. [44] investigated the association between HCV seropositivity and acute myocardial infarction using a well-established cohort of young men in the US military and found no evidence (adjusted RR, 0.94; 95 % CI, 0.52–1.68) to support this association (292 case patients and 290 control subjects). In the study of Bedimo et al. [45] including a total of 19,424 HIV-infected patients, 31.6 % of whom were HCV-coinfected, HCV coinfection was associated with lower cholesterol levels but with higher rates of hypertension, acute myocardial infarction, and cardiovascular diseases in total (OR = 1.25 with 95 % CI = 0.98–1.61 for acute myocardial infarction and 1.20 with 95 % CI = 1.04–1.38 for cardiovascular diseases). However, D:A:D Cohort Study did not find any association between HCV coinfection and the development of myocardial infarction among HIV-infected individuals [27]. Event rates per 1,000 person-years in those who were HCV-seronegative and HCV-seropositive were 3.3 (95 % CI = 3.0–3.7) and 2.7 (95 % CI = 2.2–3.3), respectively [27]. After adjustment, there was no association between HCV seropositivity and the development of myocardial infarction (rate ratio 0.86, 95 % CI = 0.62–1.19) [27].

5.3.5   HCV and Stroke

Liao et al. [46] evaluated the risk of stroke in association with chronic HCV infection in a longitudinal population-based cohort study including 4,094 adults newly diagnosed with HCV infection and 16,376 adults without HCV infection. The cumulative risk of stroke for people with HCV and without HCV infection was 2.5 % and 1.9 %, respectively, and compared to people without HCV infection, the adjusted hazard ratio of stroke was 1.27 (95 % CI = 1.14–1.41) for people with HCV infection [46]. In community-based prospective cohort study of Lee et al. [47], the cumulative risk of cerebrovascular deaths was 1.0 % and 2.7 % for seronegatives and seropositives of anti-HCV, respectively. The hazard ratio of cerebrovascular death was 2.18 (95 % CI = 1.50–3.16) for anti-HCV seropositives after adjustment for several conventional risk factors of cerebrovascular disease. Compared to participants seronegative for anti-HCV, the multivariate-adjusted hazard ratio was 1.4 (95 % CI = 0.62–3.16), 2.36 (95 % CI = 1.42–3.93), and 2.82 (95 % CI = 1.25–6.37), respectively, for anti-HCV-seropositive participants with undetectable, low, and high serum levels of HCV RNA. However, no significant association was observed between HCV genotype and cerebrovascular death [47].

5.3.6   Discussion

It seems to be that chronic HCV infection is associated with subclinical atherosclerosis, CAD, and stroke, but its correlation with myocardial infarction is arguable. For subclinical atherosclerosis, the association may be population-dependent (positive for Japanese, Italian, and French populations but non-significant in US population), whereas all studies investigating the role of HCV in the development of CAD or stroke obtained only positive results. A number of studies suggested that chronic HCV infection is associated with favorable lipid profile [28, 39, 40, 43], namely, with elevated HDL-C and reduced TC, LDL-C and TGs, so the metabolic mechanism of atherosclerosis development does not play a significant role in the case with chronic HCV infection. In contrast, significant proinflammatory alterations were noted during chronic HCV infection including high levels of TNF-α, IL-6, sVCAM-1, sICAM-1 [31, 32, 42]. However, the results of these findings may also be contradicted since there were found that HCV chronically infected patients are characterized by a lower level of CRP and fibrinogen. Nevertheless, the fact that HCV RNA sequences were detected in plaque tissue supports the hypothesis about HCV-caused active local infection in blood vessels [29]. In light of this discovery, the inflammatory mechanisms of causing atherosclerosis seem to be at least probable. In addition, HCV seropositivity was found to be associated with an increased risk of carotid-artery plaque (OR = 1.92, 95 % CI 1.56–2.38) and carotid intima-media thickening (OR = 2.85, 95 % CI = 2.28–3.57) that may reflect HCV participation in the pathogenesis of carotid arterial remodeling [48]. It is also worth a note that HCV may affect the gene expression in the endothelial cells or fibroblasts promoting the plaque formation. So, it seems to be that HCV is the only virus possibly affecting atherosclerosis development directly. A significant number of studies carried out with positive result also suggests that chronic HCV infection is associated with various cardiovascular pathologies including subclinical atherosclerosis, CAD, and stroke, although its influence on the development of distinct types of stroke (ischaemic and hemorrhagic) remains obscure.

5.4 Conclusions

All epidemiological studies devoted to the association of hepatitis viruses with atherosclerosis and related diseases are summarized in Table 5.1. To sum up, it is hard to conclude definitely whether HAV is associated with atherosclerosis and related diseases but it seems to be that this correlation is population-dependent, and may be based on the causation of endothelial dysfunction in terms of decrease of acetylcholine-mediated microvascular and epicardial dilation by this virus. Regarding HBV, the reliable association of the chronic infection caused by this virus with atherosclerosis and related diseases was not revealed, although it is possible that it decreases the risk of myocardial infarction and ischaemic stroke whilst increasing the risk of hemorrhagic stroke via the reduction of synthesis of coagulation factors in liver affected by infection. Finally, it was demonstrated that chronic HCV infection is associated with subclinical atherosclerosis, CAD, and stroke, most probably through the direct inflammatory mechanism. According to this hypothesis, HCV may affect blood vessels, feasibly initiating and promoting the plaque formation. In addition, indirect inflammatory mechanism taking a place in the affected liver is also plausible. However, it does not seem to be that hepatitis viruses are promoting atherosclerosis through atherotic mechanisms since chronic HBV and HCV infection are associated with favorable lipid pattern (decrease of TCs, TGs, LDL-C, and increase of HDL-C) which probably occurs due to chronic hepatitis and, in some cases, liver cirrhosis. Further basic investigations devoted to the basic mechanisms of development of hepatitis viruses-related atherosclerosis are needed, and they should shed light on this problem.

Table 5.1 Association of HAV, HBV, and HCV with risk of atherosclerosis and related diseases in epidemiological studies

Author, reference, population	Year	Sample size	Association, OR (95 % CI)
HAV			
Zhu et al. [1] US population	2000	391 patients undergoing coronary angiography (52.4 % with anti-HAV IgG)	2.3 (1.33–4.03) for subclinical atherosclerosis
Zhu et al. [2] US population	2001	890 patients with CAD (53.4 % with anti-HAV IgG)	1.6 (1.1–2.3) for myocardial infarction
Smieja et al. [11] Canadian population	2001	170 patients with myocardial infarction or unstable angina (59.2 % with anti-HAV IgG) and 107 controls	No association with cardiovascular events
Auer et al. [13] Austrian population	2003	218 patients undergoing coronary angiography (81.7 % with anti-HAV IgG)	No association with CAD
Ongey et al. [14] German population	2004	365 patients with diabetes mellitus from 4,285 participants of the German Health Survey	No association with cardiovascular diseases
Smieja et al. [15] Canadian population	2003	3,168 patients (76 % with anti-HAV IgG) from the Heart Outcomes Prevention Evaluation (HOPE) Study	No association with cardiovascular events
HBV			
Subclinical atherosclerosis			
Ishizaka et al. [16] Japanese population	2002	4,686 study subjects (0.9 % HBsAg-positive)	1.57 (1.10–2.24)
Bilora et al. [17] Italian population	2002	48 patients and 50 controls	No association
Su et al. [19] Taiwanese population	2004	1,330 subjects	No association
Yang et al. [22] Taiwanese population	2007	508 subjects (17.1 % HBsAg-positive)	No association
CAD			
Tong et al. [20] Chinese population	2005	291 patients with CAD (77 % HBV-positive) and 143 controls (73.4 % HBV-positive)	No association

Study / population	Year	Subjects	Result
Ghotaslou et al. [21] Iranian population	2008	4,499 patients with CAD (3.3 % HBsAg-positive) and 505 controls (2.2 % HBsAg-positive)	No association
Grub et al. [26] Norwegian population	2012	67 patients with inflammatory rheumatic diseases (IRD), 52 patients without IRD and 30 healthy controls	No association
Myocardial infarction and stroke			
Völzke et al. [18] German population	2004	4,033 subjects (5 % HBsAg-positive)	No association
Sung et al. [23] Korean population	2007	521,421 individuals	0.79 (0.68–0.90) for ischaemic stroke 0.74 (0.62–0.87) for myocardial infarction 1.33 (1.15–1.52) for hemorrhagic stroke
Wang et al. [25] Taiwanese population	2010	22,472 subjects (17.5 % HBsAg-positive)	No association
Data Collection on Adverse Events of Anti-HIV Drugs (D:A:D) Study Group [27]	2010	33,347 individuals	No association
HCV			
Subclinical atherosclerosis			
Ishizaka et al. [34] Japanese population	2003	1,992 subjects (1.3 % HCV-seropositive)	5.61 (2.06–15.26)
Targher et al. [35] Italian population	2007	60 consecutive patients with biopsy-proven non-alcoholic steatohepatitis, 60 patients with HCV, 35 patients with HBV, and 60 healthy controls	Increased risk
Caliskan et al. [37] Turkish population	2008	72 chronic hemodialysis patients (50 % HCV-positive)	No association
Tien et al. [38] US population	2010	1,223 HIV-infected subjects (22.3 % HCV-positive), 452 controls	No association
Miyajima et al. [39] Japanese population	2012	40 subjects with chronic HCV infection, 88 individuals with transient HCV infection, 1,780 controls	Increased risk

(continued)

Table 5.1 (continued)

Author, reference, population	Year	Sample size	Association, OR (95 % CI)
Sosner et al. [40] French population	2012	18 HCV-HIV co-infected patients and 22 HIV mono-infected patients	10 (1.5–72)
CAD			
Vassalle et al. [41] Italian population	2004	491 patients and 195 controls	4.2 (1.4–13)
Tsui et al. [42] US population	2009	971 patients (8.65 % HCV-positive)	2.25 (1.02–4.97) for heart failure
Butt et al. [43] US population	2009	82,083 HCV-infected and 89,582 HCV-uninfected subjects	1.25 (1.2–1.3)
Myocardial infarction			
Arcari et al. [44] US population	2006	292 cases and 290 controls	No association
Bedimo et al. [45] US population	2010	19,424 HIV-infected patients (31.6 % HCV-positive)	1.25 (0.98–1.61)
Stroke			
Liao et al. [46] Taiwanese population	2012	4,094 HCV-positive and 16,376 HCV-negative subjects	1.27 (1.14–1.41)
Lee et al. [47] Taiwanese population	2010	23,665 subjects (5.5 % HCV-positive)	1.40 (0.62–3.16) for anti-HCV-seropositive participants with undetectable serum level of HCV RNA 2.36 (1.42–3.93) for anti-HCV-seropositive participants with low serum level of HCV RNA 2.82 (1.25–6.37) for anti-HCV-seropositive participants with high serum level of HCV RNA

References

1. Zhu J, Quyyumi AA, Norman JE, Costello R, Csako G, Epstein SE. The possible role of hepatitis A virus in the pathogenesis of atherosclerosis. J Infect Dis. 2000;182:1583–7.
2. Zhu J, Nieto FJ, Horne BD, Anderson JL, Muhlestein JB, Epstein SE. Prospective study of pathogen burden and risk of myocardial infarction or death. Circulation. 2001;103:45–51.
3. Prasad A, Zhu J, Halcox JP, Waclawiw MA, Epstein SE, Quyyumi AA. Predisposition to atherosclerosis by infections: role of endothelial dysfunction. Circulation. 2002;106:184–90.
4. Cainelli F, Concia E, Vento S. Hepatitis A virus infection and atherosclerosis. J Infect Dis. 2001;184:390–1.
5. Ramia S. Antibody against hepatitis A in Saudi Arabians and in expatriates from various parts of the world working in Saudi Arabia. J Infect. 1986;12:153–5.
6. Darwish MA, Faris R, Clemens JD, Rao MR, Edelman R. High seroprevalence of hepatitis A, B, C, and E viruses in residents in an Egyptian village in the Nile delta: a pilot study. Am J Trop Med Hyg. 1996;54:554–8.
7. Mele A, Pasquini P, Pana A. Hepatitis A in Italy: epidemiology and suggestions for control. Ital J Gastroenterol. 1991;23:341–3.
8. Amin J, Heath T, Morrell S. Hepatitis A in Australia in the 1990s: future directions in surveillance and control. Commun Dis Intell. 1999;23:113–20.
9. Massaro M, Carluccio MA, De Caterina R. Direct vascular antiatherogenic effects of oleic acid: a clue to the cardioprotective effects of the Mediterranean diet. Cardiologia. 1999;44:507–13.
10. Pearson TA. Cardiovascular disease in developing countries: myths, realities, and opportunities. Cardiovasc Drugs Ther. 1999;13:95–104.
11. Smieja M, Cronin L, Levine M, Goldsmith CH, Yusuf S, Mahony JB. Previous exposure to Chlamydia pneumoniae, Helicobacter pylori and other infections in Canadian patients with ischemic heart disease. Can J Cardiol. 2001;17:270–6.
12. Burnett MS, Zhu J, Miller JM, Epstein SE. Effects of hepatitis A vaccination on atherogenesis in a murine model. J Viral Hepat. 2003;10:433–6.
13. Auer J, Leitinger M, Berent R, Prammer W, Weber T, Lassnig E, Eber B. Hepatitis A IgG seropositivity and coronary atherosclerosis assessed by angiography. Int J Cardiol. 2003;90:175–9.
14. Ongey M, Brenner H, Thefeld W, Rothenbacher D. Helicobacter pylori and hepatitis A virus infections and the cardiovascular risk profile in patients with diabetes mellitus: results of a population-based study. Eur J Cardiovasc Prev Rehabil. 2004;11:471–6.
15. Smieja M, Gnarpe J, Lonn E, Gnarpe H, Olsson G, Yi Q, Dzavik V, McQueen M, Yusuf S, Heart Outcomes Prevention Evaluation (HOPE) Study Investigators. Multiple infections and subsequent cardiovascular events in the Heart Outcomes Prevention Evaluation (HOPE) Study. Circulation. 2003;107:251–7.
16. Ishizaka N, Ishizaka Y, Takahashi E, Toda Ei E, Hashimoto H, Ohno M, Nagai R, Yamakado M. Increased prevalence of carotid atherosclerosis in hepatitis B virus carriers. Circulation. 2002;105:1028–30.
17. Bilora F, Rinaldi R, Boccioletti V, Petrobelli F, Girolami A. Chronic viral hepatitis: a prospective factor against atherosclerosis. A study with echo-color Doppler of the carotid and femoral arteries and the abdominal aorta. Gastroenterol Clin Biol. 2002;26:1001–4.
18. Völzke H, Schwahn C, Wolff B, Mentel R, Robinson DM, Kleine V, Felix SB, John U. Hepatitis B and C virus infection and the risk of atherosclerosis in a general population. Atherosclerosis. 2004;174:99–103.
19. Su TC, Lee YT, Cheng TJ, Chien HP, Wang JD. Chronic hepatitis B virus infection and dyslipidemia. J Formos Med Assoc. 2004;103:286–91.
20. Tong DY, Wang XH, Xu CF, Yang YZ, Xiong SD. Hepatitis B virus infection and coronary atherosclerosis: results from a population with relatively high prevalence of hepatitis B virus. World J Gastroenterol. 2005;11:1292–6.

21. Ghotaslou R, Aslanabadi N, Ghojazadeh M. Hepatitis B virus infection and the risk of coronary atherosclerosis. Ann Acad Med Singapore. 2008;37:913–5.
22. Yang KC, Chen MF, Su TC, Jeng JS, Hwang BS, Lin LY, Liau CS, Lee YT. Hepatitis B virus seropositivity is not associated with increased risk of carotid atherosclerosis in Taiwanese. Atherosclerosis. 2007;195:392–7.
23. Sung J, Song YM, Choi YH, Ebrahim S, Davey Smith G. Hepatitis B virus seropositivity and the risk of stroke and myocardial infarction. Stroke. 2007;38:1436–41.
24. Turhan O, Coban E, Inan D, Yalcin AN. Increased mean platelet volume in chronic hepatitis B patients with inactive disease. Med Sci Monit. 2010;16:CR202–5.
25. Wang CH, Chen CJ, Lee MH, Yang HI, Hsiao CK. Chronic hepatitis B infection and risk of atherosclerosis-related mortality: a 17-year follow-up study based on 22,472 residents in Taiwan. Atherosclerosis. 2010;211:624–9.
26. Grub C, Brunborg C, Hasseltvedt V, Aukrust P, Førre O, Almdahl SM, Hollan I. Antibodies to common infectious agents in coronary artery disease patients with and without rheumatic conditions. Rheumatology (Oxford). 2012;51:679–85.
27. Data Collection on Adverse Events of Anti-HIV Drugs (D:A:D) Study Group, Weber R, Sabin C, Reiss P, de Wit S, Worm SW, Law M, Dabis F, D'Arminio Monforte A, Fontas E, El-Sadr W, Kirk O, Rickenbach M, Phillips A, Ledergerber B, Lundgren J. HBV or HCV coinfections and risk of myocardial infarction in HIV-infected individuals: the D:A:D Cohort Study. Antivir Ther. 2010;15:1077–86.
28. Marzouk D, Sass J, Bakr I, El Hosseiny M, Abdel-Hamid M, Rekacewicz C, Chaturvedi N, Mohamed MK, Fontanet A. Metabolic and cardiovascular risk profiles and hepatitis C virus infection in rural Egypt. Gut. 2007;56:1105–10.
29. Boddi M, Abbate R, Chellini B, Giusti B, Giannini C, Pratesi G, Rossi L, Pratesi C, Gensini GF, Paperetti L, Zignego AL. Hepatitis C virus RNA localization in human carotid plaques. J Clin Virol. 2010;47:72–5.
30. Masiá M, Robledano C, López N, Escolano C, Gutiérrez F. Treatment for hepatitis C virus with pegylated interferon-α plus ribavirin induces anti-atherogenic effects oncardiovascular risk biomarkers in HIV-infected and -uninfected patients. J Antimicrob Chemother. 2011;66: 1861–8.
31. Masiá M, Padilla S, Robledano C, Ramos JM, Gutiérrez F. Evaluation of endothelial function and subclinical atherosclerosis in association with hepatitis C virus in HIV-infected patients: a cross-sectional study. BMC Infect Dis. 2011;11:265.
32. Oliveira CP, Kappel CR, Siqueira ER, Lima VM, Stefano JT, Michalczuk MT, Marini SS, Barbeiro HV, Soriano FG, Carrilho FJ, Pereira LM, Alvares-da-Silva MR. Effects of Hepatitis C virus on cardiovascular risk in infected patients: a comparative study. Int J Cardiol. 2013;164(2):221–6.
33. Perrin-Cocon L, Diaz O, André P, Lotteau V. Modified lipoproteins provide lipids that modulate dendritic cell immune function. Biochimie. 2013;95(1):103–8.
34. Ishizaka Y, Ishizaka N, Takahashi E, Unuma T, Tooda E, Hashimoto H, Nagai R, Yamakado M. Association between hepatitis C virus core protein and carotid atherosclerosis. Circ J. 2003;67:26–30.
35. Targher G, Bertolini L, Padovani R, Rodella S, Arcaro G, Day C. Differences and similarities in early atherosclerosis between patients with non-alcoholic steatohepatitis and chronic hepatitis B and C. J Hepatol. 2007;46:1126–32.
36. Sawayama Y, Okada K, Maeda S, Ohnishi H, Furusyo N, Hayashi J. Both hepatitis C virus and Chlamydia pneumoniae infection are related to the progression of carotid atherosclerosis in patients undergoing lipid lowering therapy. Fukuoka Igaku Zasshi. 2006;97:245–55.
37. Caliskan Y, Oflaz H, Pusuroglu H, Boz H, Yazici H, Tamer S, Karsidag K, Yildiz A. Hepatitis C virus infection in hemodialysis patients is not associated with insulin resistance, inflammation and atherosclerosis. Clin Nephrol. 2009;71:147–57.
38. Tien PC, Schneider MF, Cole SR, Cohen MH, Glesby MJ, Lazar J, Young M, Mack W, Hodis HN, Kaplan RC. Association of hepatitis C virus and HIV infection with subclinical atherosclerosis in the women's interagency HIV study. AIDS. 2009;23:1781–4.

39. Miyajima I, Kawaguchi T, Fukami A, Nagao Y, Adachi H, Sasaki S, Imaizumi T, Sata M. Chronic HCV infection was associated with severe insulin resistance and mild atherosclerosis: a population-based study in an HCV hyperendemic area. J Gastroenterol. 2013;48(1):93–100.

40. Sosner P, Wangermez M, Chagneau-Derrode C, Le Moal G, Silvain C. Atherosclerosis risk in HIV-infected patients: the influence of hepatitis C virus co-infection. Atherosclerosis. 2012; 222:274–7.

41. Vassalle C, Masini S, Bianchi F, Zucchelli GC. Evidence for association between hepatitis C virus seropositivity and coronary artery disease. Heart. 2004;90:565–6.

42. Tsui JI, Whooley MA, Monto A, Seal K, Tien PC, Shlipak M. Association of hepatitis C virus seropositivity with inflammatory markers and heart failure in persons with coronary heart disease: data from the Heart and Soul study. J Card Fail. 2009;15:451–6.

43. Butt AA, Xiaoqiang W, Budoff M, Leaf D, Kuller LH, Justice AC. Hepatitis C virus infection and the risk of coronary disease. Clin Infect Dis. 2009;49:225–32.

44. Arcari CM, Nelson KE, Netski DM, Nieto FJ, Gaydos CA. No association between hepatitis C virus seropositivity and acute myocardial infarction. Clin Infect Dis. 2006;43:e53–6.

45. Bedimo R, Westfall AO, Mugavero M, Drechsler H, Khanna N, Saag M. Hepatitis C virus coinfection and the risk of cardiovascular disease among HIV-infected patients. HIV Med. 2010;11:462–8.

46. Liao CC, Su TC, Sung FC, Chou WH, Chen TL. Does hepatitis C virus infection increase risk for stroke? A population-based cohort study. PLoS One. 2012;7:e31527.

47. Lee MH, Yang HI, Wang CH, Jen CL, Yeh SH, Liu CJ, You SL, Chen WJ, Chen CJ. Hepatitis C virus infection and increased risk of cerebrovascular disease. Stroke. 2010;41:2894–900.

48. Ishizaka N, Ishizaka Y, Takahashi E, Tooda E, Hashimoto H, Nagai R, Yamakado M. Association between hepatitis C virus seropositivity, carotid-artery plaque, and intima-media thickening. Lancet. 2002;359:133–5.

Chapter 6
Summary: Are We There Yet?

Abstract In this chapter, we will summarize the data obtained from the analysis in previous chapters. We will define the criteria for the credibility of the connection of the viruses considered in this review with atherosclerosis and related diseases and will consider the principles of organization of studies in this field.

In general, the criteria for the credibility of the connection of the viruses considered in this review with atherosclerosis and related diseases may be postulated as follows:

- Viruses should be detected in the significant share of plaques from atherosclerosis cases (enterovirus, parvovirus B19, EBV, HCV);
- Viruses must be localized in atherosclerotic tissues significantly more frequently and/or in higher concentration compared to corresponding tissues collected from healthy controls (enterovirus, EBV);
- Viruses must colonize the tissue before atherosclerosis development, at every stage of plaque atherogenesis or even when tissues are absolutely normal (unknown for any of the indicated viruses);
- Viruses must have eminent plaque formation activity on the molecular and cellular level (it must have an ability to cause immune response and promote atherogenesis). They may cause mutations leading to the accumulation of lymphocytes in the plaque, inhibit their apoptosis, enhance their growth, accelerate their proliferation, support their survival, and promote plaque stroma restructuring. This statement may be true for enterovirus and EBV, according to the published data;
- If viruses do not possess the direct plaque-forming effect they should possess indirect effects (e.g. causing chronic systemic inflammation, promoting a development of unfavorable lipid profile or affecting coagulation ability leading to atherothrombosis or hemorrhage). It may be true for HBV, since chronic HBV infection may be associated with elevated risk of hemorrhagic stroke via decrease of coagulation ability due to liver affection, and also for EBV, which may cause

alterations in lipid profile making it unfavorable; indirect systemic inflammation may be also caused by HBV and HCV causing chronic hepatitis;
- An association between viruses and atherosclerosis must be demonstrated in animal models by different research groups (unknown for any of the indicated viruses);
- Etiotropic therapy should lead to the regression of atherosclerotic plaque associated with viral infection (unknown for any of the indicated viruses, possibly excepting HCV);
- The association between viruses and atherosclerosis risk must be confirmed in large prospective and retrospective well-designed epidemiological studies in various countries and desirably by different research groups (enterovirus, with some limitations, and HCV);

For the credible determination of the possible association of viruses with atherosclerosis and related diseases, the following principles of study organization can be postulated:

- Investigation of the immune, atherogenic, or coagulative mechanisms of the viral impact on the development of atherosclerosis and related diseases;
- The interactions between virus-related mechanisms and other mechanisms of atherosclerosis etiopathogenesis can be investigated;
- It is important to use the most sensitive and specific markers of viral infection (for instance, PCR or serological tests; in the latter case the diagnostic titer of relevant antibodies should be the most sensitive and specific one);
- In case if it is possible, the investigations should be conducted in various populations to reveal a feasible population dependency of the association (to reach this aim, international consortia may be created)

It is also worth of note that there are a lot of projects devoted to a development of vaccines against viruses that are described in this book. In the case if these vaccines will be implemented, it would be possible to investigate their role in decreasing incidence of atherosclerosis and related diseases. It may be an additional way to reveal an association of these viruses with cardiovascular disorders. A vaccine against influenza which is actively used worldwide might be a good example. In certain investigations, it was shown that vaccinated subjects are less prone to have acute cardiovascular events compared to non-vaccinated (Rogers et al. [1], Grau et al. [2], Warren-Gash et al. [3], Lang et al. [4]); however, it is difficult to differentiate whether influenza virus is a true cause of atherosclerosis, or it just causes an acute infection and intoxication that may significantly disrupt heart functioning, particularly in elderly. Undoubtedly, future studies will shed light on the problem of the role of viruses in atherosclerosis and related diseases.

References

1. Rogers KC, Wallace JP, Foster SL, Finks SW. Annual influenza vaccination: offering protection beyond infection. South Med J. 2012;105:379–86.
2. Grau AJ, Urbanek C, Palm F. Common infections and the risk of stroke. Nat Rev Neurol. 2010;6:681–94.

3. Warren-Gash C, Smeeth L, Hayward AC. Influenza as a trigger for acute myocardial infarction or death from cardiovascular disease: a systematic review. Lancet Infect Dis. 2009;9: 601–10.
4. Lang PO, Mendes A, Socquet J, Assir N, Govind S, Aspinall R. Effectiveness of influenza vaccine in aging and older adults: comprehensive analysis of the evidence. Clin Interv Aging. 2012;7:55–64.

Index

A. Kutikhin et al., *Viruses and Atherosclerosis*, SpringerBriefs in Immunology 4,
DOI 10.1007/978-1-4614-8863-7, © Springer Science+Business Media New York 2013